T0353408

Progress in Scientific Computing
Vol. 5

Edited by
S. Abarbanel
R. Glowinski
G. Golub
P. Henrici
H.-O. Kreiss

Birkhäuser
Boston · Basel · Stuttgart

Numerical Boundary Value ODEs

Proceedings of an International Workshop, Vancouver, Canada, July 10–13, 1984

U. M. Ascher
R. D. Russell, editors

1985

Birkhäuser
Boston · Basel · Stuttgart

Uri M. Ascher
The University of British Columbia
Vancouver, B.C. V6T 1W5
(Canada)

Robert D. Russell
University of New Mexico
Albuquerque, NM 87131
(USA)

Library of Congress Cataloging in Publication Data
Main entry under title:

Numerical boundary value ODEs.

(Progress in scientific computing ; v. 5)
1. Boundary value problems -- Numerical solutions -- Congresses. 2. Differential equations -- Numerical solutions -- Congresses. I. Ascher, U. M. (Uri M.), 1946– . II. Russell, R. D. (Robert D.), 1945– III. Series.
QA379.N86 1985 515.3'5 85–6234
ISBN 0-8176-3302-2

CIP-Kurztitelaufnahme der Deutschen Bibliothek

Numerical boundary value ODEs : proceedings of an internat. workshop, Vancouver, Canada, July 10–13, 1984 / U. M. Ascher ; R. D. Russell, ed. – Boston ; Basel ; Stuttgart : Birkhäuser, 1985.
(Progress in scientific computing ; Vol. 5)
ISBN 3-7643-3302-2 (Stuttgart . . .)
ISBN 0-8176-3302-2 (Boston)

NE: Ascher, Uri M. [Hrsg.]; GT

Printed in Germany
ISBN 0-8176-3302-2
ISBN 3-7643-3302-2

CONTENTS

PREFACE

In the past few years, knowledge about methods for the numerical solution of two-point boundary value problems has increased significantly. Important theoretical and practical advances have been made in a number of fronts, although they are not adequately described in any text currently available. With this in mind, we organized an international workshop, devoted solely to this topic. The workshop took place in Vancouver, B.C., Canada, in July 10-13, 1984. This volume contains the *refereed* proceedings of the workshop.

Contributions to the workshop were in two formats. There were a small number of invited talks (ten of which are presented in this proceedings); the other contributions were in the form of poster sessions, for which there was no parallel activity in the workshop. We had attempted to cover a number of topics and objectives in the talks. As a result, the general review papers of O'Malley and Russell are intended to take a broader perspective, while the other papers are more specific.

The contributions in this volume are divided (somewhat arbitrarily) into five groups. The first group concerns fundamental issues like conditioning and decoupling, which have only recently gained a proper appreciation of their centrality. Understanding of certain aspects of shooting methods ties in with these fundamental concepts. The papers of Russell, de Hoog and Mattheij all deal with these issues.

The second group of papers is concerned with various implementation aspects for several methods. Enright's paper presents a general view and deals with Runge-Kutta methods. The third group considers the solution of stiff boundary value problems. The survey of O'Malley covers analytical and numerical aspects of solving singularly perturbed problems, while Ascher's paper is about symmetric difference schemes.

In the fourth group we have combined two extensions. Bader's paper is on the numerical solution of functional (dealy) differential equations, while Griewank and Reddien deal with problems of continuation and bifurcation. Finally, in the fifth group specific applications are discussed: Markowich considers semiconductors, while Smooke considers an application in combustion theory.

Additional contributions in the first four categories by van Loon, England-Mattheij, Krogh et al., Sleptsov, Brown, Maier, Bellen and Seydel are also included in this proceedings.

The participants in the workshop were a most congenial group, and their relaxed, frank attitude helped to make this event both informal and successful.

In organizing the workshop we were aided by Dr. Brian Seymour, who was the local organizer of the Canadian Applied Mathematics Society meeting, which was held concurrently with ours at the University of British Columbia. Thanks are also due to the University Conference Centre and, especially, to the staff of the Computer Science Department for their help.

Finally, let us gratefully acknowledge the support we have received from the Canadian National Science and Engineering Research Council (grants 650-709, A4306 and A8781). Without this support, the workshop and this proceedings could not have been made possible.

February 1985

Uri Ascher and Robert D. Russell

LIST OF PARTICIPANTS

Dr. Clifford Addison
Computer Science Department
University of Alberta
Edmonton, Alberta T6G 2H1
(Canada)

Dr. Uri Ascher
Computer Science Department
University of British Columbia
Vancouver, B.C. V6T 1W5
(Canada)

Dr. Georg Bader
Mathematics Department
Simon Fraser University
Burnaby, B.C. V5A 1S6
(Canada)

Dr. Ivo Babuska
Inst. for Physical Science & Techn.
University of Maryland
College Park, MD 20742 (USA)

Dr. Paul Bailey
Numerical Mathematics Div. - 1642
Sandia National Laboratories
Albuquerque, NM 87185 (USA)

Dr. Alfredo Bellen
Istituto di Matematica
Università degli Studii
I-34100 Trieste (Italy)

Dr. David Billingsley
IBM Corporation
P.O. Box 1369
Houston, TX 77251 (USA)

Dr. David Brown
Center for Nonlinear Studies
MS-B258
Los Alamos National Labs
Los Alamos, NM 87545 (USA)

Dr. Jeff Cash
Mathematics Department
Imperial College
South Kensington
London S.W.7 (England)

Dr. Ray C.Y. Chin
Lawrence National Laboratory
P.O. Box 808, L-71
Livermore, CA 94550 (USA)

Mr. Robert Corless
Mechanical Engineering
University of British Columbia
Vancouver, B.C. V6T 1W5
(Canada)

Dr. Julio Diaz
Computer Science Department
University of Oklahoma
Norman, OK 73019 (USA)

Mr. Luca Dieci
Mathematics Department
University of New Mexiko
Albuquerque, NM 87131 (USA)

Dr. Dugald Duncan
Atomic Energy of Canada
Chalk River, Ont. (Canada)

Dr. Ronald England
IIMAS-UNAM
Apdo. Postal 20-726
Mexico D.F. (Mexico)

Dr. Wayne Enright
Computer Science Department
University of Toronto
Toronto, Ont. M5S 1A7
(Canada)

Dr. Graeme Fairweather
Mathematics Department
University of Kentucky
Lexington, KY 40506 (USA)

Dr. Ian Gladwell
Mathematics Department
University of Manchester
Manchester M13 9PL (England)

Dr. Andreas Griewank
Mathematics Department
Southern Methodist University
Dallas, TX 75275 (USA)

Dr. Suchitra Gupta
Computer Science Department
Pennsylvania State University
University Park, PA 16802 (USA)

Dr. Patricia Hanson
Computer Science Department
Edmonton, Alb. T6G 2H1 (Canada)

Dr. Richard Hickman
Lawrence Livermore Laboratory
P.O. Box 808
Livermore, CA 94550 (USA)

Dr. Frank de Hoog
CSIRO
Division of Mathematics and Statistics
P.O. Box 1965
Canberra, ACT 2601 (Australia)

Dr. Fred Howes
Mathematics Department
University of California at Davis
Davis, CA 95616 (USA)

Dr. Laurie Johnston
Computer Science Department
University of Toronto
Toronto, Ont. M5S 1A7 (Canada)

Dr. Mohan Kadalbajoo
Mathematics Department
Indian Institute of Technology
Kanpur-208016 (India)

Dr. Chittaranjan Katti
M-64 Connaught Circus
New Delhi-110001 (India)

Dr. Baker Kearfott
Computing Technology and Services Division
Exxon Research and Engineering Co.
Clinton Township, Route 22 East
Annandale, NJ 08801 (USA)

Dr. Pat Keast
Department of Mathematics, Statistics and Computer Science
Dalhousie University
Halifax, Nova Scotia B3H 4H8 (Canada)

Dr. Herb Keller
Department of Applied Mathematics
California Institute of Technology
Pasadena, CA 91125 (USA)

Dr. Gregory Kriegsman
Mathematics Department
Northwestern University
Evanston, IL 60201 (USA)

Dr. Fred Krogh
Jet Propulsion Laboratory
4800 Oak Grove Drive
Pasadena, CA 91103 (USA)

Dr. Chuck Lange
Mathematics Department
UCLA
Los Angeles, CA 90024 (USA)

Dr. John Lavery
Department of Mathematics and Statistics
Case Western Reserve University
Cleveland, OH 44106 (USA)

Dr. J. Doug Lawson
Computer Science Department
University of Waterloo
Waterloo, Ont. N2L 3G1 (Canada)

Mr. Richard Lee
TRIUMF
University of British Columbia
Vancouver, B.C. V6T 1W5 (Canada)

Dr. Marianela Lentini
Depto. de Matematicas
Universidad Simon Bolivar
Apdo. 80659
Caracas 1081 (Venezuela)

Dr. Gil Lewis
Mathematics Department
Michigan Technological University
Houghton, MI 49855 (USA)

Dr. Deborah Lockhart
Mathematics Department
Michigan Technological University
Houghton, MI 49855 (USA)

Dr. Paul van Loon
Computing Centre
Eindhoven University of Technology
P.O. Box 513
5600 MB Eindhoven
(The Netherlands)

Dr. Max Maier
Mathematisches Institut der
Technischen Universität München
Arcisstrasse 21
Postfach 20 24 20
D-8000 München 2 (FRG)

Dr. Vaclav Majer
Institute for Physical Science
and Technology
College Park, MD 20742 (USA)

Dr. Peter Markowich
Institut für Angewandte und
Numerische Mathematik
Technische Universität Wien
A-1040 Wien (Austria)

Dr. Bob Mattheij
Mathematisch Institut
Katholieke Universiteit
Nijmegen (The Netherlands)

Dr. Derek Meek
Computer Science Department
University of Manitoba
Winnipeg, Man. R3T 2N2
(Canada)

Dr. Robert Miura
Mathematics Department
University of British Columbia
Vancouver, B.C. V6T 1W5
(Canada)

Dr. Paul Muir
Computer Science Department
University of Toronto
Toronto, Ont. M5S 1A7
(Canada)

Dr. Chris Newberry
Mathematics Department
University of Kentucky
Lexington, KY 40506 (USA)

Dr. Viatseslav Novikov
Institute for Theoretical and
Applied Mechanics
Novosibirsk 630090 (USSR)

Dr. Takeo Ojika
Department of Technology
Osakakyoiku University
4-88 Minamikawahori-Cho
Tennoji-ku
Osaka (Japan)

Dr. Bob O'Malley
Mathematics Department
Rensselaer Polytechnic Institute
Troy, NY 12181 (USA)

Dr. John Paine
Mathematics Department
Simon Fraser University
Burnaby, B.C. V5A 1S6
(Canada)

Dr. Victor Pereyra
Weidlinger Association
620 Hansen Way
Palo Alto, CA 94304 (USA)

Dr. Bob Russell
Mathematics Department
Simon Fraser University
Burnaby, B.C. V5A 1S6
(Canada)

Dr. Rüdiger Seydel
Mathematics Department
SUNY at Buffalo
106 Diefendorf Hall
Buffalo, NY 14214-3093 (USA)

Dr. Philip Sharp
Computer Science Department
University of Toronto
Toronto, Ont. M5S 1A7
(Canada)

Dr. Bruce Simpson
Computer Science Department
University of Waterloo
Waterloo, Ont. N2L 3G1
(Canada)

Dr. Anatol Sleptsov
Institute for Theoretical and
Applied Mechanics
Novosibirsk 630090 (USSR)

Dr. Mitch Smooke
Mechanical Engineering Department
Yale University
New Haven, CT 06520 (USA)

Dr. Frank Stenger
Mathematics Department
University of Utah
Salt Lake City, UT 84112 (USA)

Dr. Paul Swarztrauber
National Center for Atmospheric Research
P.O. Box 3000
Boulder, CO 80307 (USA)

Dr. Reginald Tewarson
Department of Applied Mathematics and Statistics
SUNY at Stony Brook
Long Island, NY 11794 (USA)

Dr. Lewis Thigpen
Lawrence Livermore National Laboratory
P.O. Box 808, L-71
Livermore, CA 94550 (USA)

Dr. Per Grove Thomsen
Numerical Analytical Department
Technical University of Denmark
Building 303
DK-2800 Lyngby (Denmark)

Dr. Jim Varah
Computer Science Department
University of British Columbia
Vancouver, B.C. V6T 1W5
(Canada)

Dr. Rien van Veldhuizen
Wiskundig Seminarium
Vrije Universiteit
P.O. Box 7161
NL-1007 MC Amsterdam
(The Netherlands)

Dr. Fred Wan
Applied Mathematics Program
University of Washington
Seattle, WA 98195 (USA)

Dr. Watanabe Satoshi
Tohoku Dental University

Dr. H.A. Watts
Applied Mathematics Division
2646 Sandia Labs
Albuquerque, NM 87115 (USA)

Dr. Richard Weiss
Institut für Angewandte und Numerische Mathematik
Technische Universität Wien
A-1040 Wien (Austria)

Mr. Stevan White
Mathematics Department
Simon Fraser University
Burnaby, B.C. V5A 1S6
(Canada)

Dr. Matt Yedlin
Geophysics Department
University of British Columbia
Vancouver, B.C. V6T 1W5
(Canada)

Dr. Ray Zahar
Mathematics Department
McGill University
Montreal, Que. H3A 2K6
(Canada)

Dr. Marino Zennaro
Istituto di Matematica
Università degli Studii
I-34100 Trieste (Italy)

Progress in Scientific Computing, Vol. 5
Numerical Boundary Value ODEs

A UNIFIED VIEW OF SOME RECENT DEVELOPMENTS IN THE NUMERICAL SOLUTION OF BVODES

By

Robert D. Russell

I. Introduction

At present it is not an easy task to learn the theory underlying the numerical methods for solving BVPs (boundary value problems) for ODEs. There has not been a single book published which treats the basic methods since the outstanding book by Keller [2] in 1968. There have been few meetings on this topic (although see [3,4,5,6]). One of the purposes of this paper is to briefly discuss numerical methods which have proved useful in practice.

There have recently been significant developments in the theory of these methods, and some of them are discussed here in order to provide background and motivation for many of the papers in this proceedings. A short overview such as this must inevitably be too brief in places, give an incomplete list of references, and reflect the author's biases of the moment.

In the next section, conditioning of linear BVPs and solution dichotomy are discussed. In section 3, several variants of multiple shooting are discussed, and their stability is related to the conditioning of the continuous problem. Marching algorithms are described in section 4; their relation to a special factorization of the multiple shooting matrix is described in section 5. In section 6 a more general matrix factorization due to Mattheij [1] is discussed. We briefly summarize the important points in the last section, emphasizing that the success of a numerical method hinges upon its ability to (implicitly or explicitly) effect a decoupling of the fundamental solution components of the underlying problem.

II. Conditioning of BVPs

We consider linear BVPs of the form

$$\mathbf{L}y := y' - A(x)y = f(x) \qquad 0 < x < 1 \tag{2.1}$$

$$\mathbf{B}y := B_0 y(0) + B_1 y(1) = b \tag{2.2}$$

where $y(x)$, $f(x)$, b are n-vectors and $A(x)$, B_0 , B_1 are $n \times n$ matrices. Assume that (2.1,2) has a unique solution, that $\max(||B_0||_\infty, ||B_1||_\infty) = 1$, and for convenience that $A(x)$, $f(x)$ are as smooth as is desired. While the concept of stability is important for IVPs (initial value problems) as a description of asymptotic behavior $(x \to \infty)$, sensitivity of BVPs on finite intervals is more appropriately described in terms of conditioning (just as is done for linear systems of equations). The description of problem conditioning is a major purpose in this section.

A fundamental solution matrix for (2.1) is a nonsingular $n \times n$ matrix (function) $Y(x)$ satisfying

$$\mathbf{L}Y = Y' - AY = 0 \qquad 0 < x < 1 . \tag{2.3}$$

Unique choice of Y requires n linearly independent conditions, e.g.,

$$Y(0) = I \tag{2.4}$$

(see [7] for other useful choices). Any other fundamental solution matrix has the form $\Phi(x) = Y(x)C$, where C is a nonsingular matrix. It is straightforward to show that

$$y(x) = Y(x)(\mathbf{B}Y)^{-1}b + \int_0^1 G(x,t)f(t)dt \tag{2.5}$$

where $\mathbf{B}Y = B_0 Y(0) + B_1 Y(1)$ and the Green's (matrix) function $G(x,t)$ is defined by

$$G(x,t) = \begin{cases} Y(x)[(\mathbf{B}Y)^{-1}B_0Y(0)]Y^{-1}(t) & x > t \\ -Y(x)[I-(\mathbf{B}Y)^{-1}B_0Y(0)]Y^{-1}(t) & x < t \end{cases} . \tag{2.6}$$

The BVP (2.1,2) is called ***well-conditioned*** [7,8] if

$$\beta = \sup_{0<x<1} ||Y(x)(\mathbf{B}Y)^{-1}||_\infty \tag{2.7}$$

and

$$\alpha = \sup_{0<x,t<1} ||G(x,t)|| \tag{2.8}$$

are "of reasonable size", viz. commensurate with the size of $||A(x)||$ but not $e^{||A(x)||}$.

A concept closely connected to conditioning is dichotomy: (2.1) has a ***dichotomy*** if there exists an $n \times n$ projection matrix P (i.e., $P^2 = P$) and constant κ "of reasonable size" such that

$$||Y(x)PY^{-1}(t)|| \leq \kappa \qquad x > t \tag{2.9}$$

$$||Y(x)(I-P)Y^{-1}(t)|| \leq \kappa \qquad x < t .$$

Furthermore, (2.1) has an ***exponential dichotomy*** if there also exist positive constants λ , λ' such that

$$||Y(x)PY^{-1}(t)|| \leq \kappa e^{-\lambda(x-t)} \qquad x > t \tag{2.10}$$

$$||Y(x)(I-P)Y^{-1}(t)|| \leq \kappa e^{-\lambda'(t-x)} \qquad x < t .$$

If P has rank r , then (exponential) dichotomy means that in a rough sense there are r (exponentially) increasing and $n-r$ (exponentially) decreasing fundamental solution components. Intuitively, we see that for well-conditioned BVPs boundary condition information at 0 and at 1 should control respectively the decreasing and increasing fundamental solutions. In fact, one can show [9] that κ , β of reasonable size $\Longrightarrow$ α of reasonable size. Having only κ or only β of reasonable size is insufficient to imply well-conditioning, as can be seen from simple examples.

A natural question to ask is whether or not

$$\alpha \textit{ of reasonable size } \Longrightarrow \kappa \textit{ of reasonable size.} \qquad (*)$$

In the separated boundary condition case, this is easy to see that $P := (\mathbf{B}Y)^{-1}B_0Y(0)$ is a projection: Letting $B_0 = \begin{bmatrix} B_0^1 \\ 0 \end{bmatrix}$, $B_1 = \begin{bmatrix} 0 \\ B_1^2 \end{bmatrix}$ where B_0^1 is $k \times n$ and B_1^2 is $(n-k) \times n$, then

$$P^2 = (\mathbf{B}Y)^{-1}B_0Y(0)(\mathbf{B}Y)^{-1}B_0Y(0)$$

$$= (\mathbf{B}Y)^{-1}\begin{bmatrix} B_0^1Y(0) \\ 0 \end{bmatrix} \begin{bmatrix} B_0^1Y(0) \\ B_1^2Y(1) \end{bmatrix}^{-1} B_0Y(0) = P \ ,$$

so (2.9) follows from (2.6,8) with $\kappa = \alpha$.

In fact $(*)$ holds in general [10], so any well-codnitioned BVP has a dichotomy. It is somewhat surprising that this result was only recently proven, for it is perhaps as fundamental as the standard stability result for the IVP for (2.1) or the now standard existence/uniqueness theory for BVPs due largely to Keller [2]. The result motivates many of the articles in this proceedings.

Since boundary conditions (BC) play such a clear role in determining decreasing (or increasing) fundamental solution components, one is led to wonder if

$$\alpha \textit{ of reasonable size } \Longrightarrow \beta \textit{ of reasonable size} \ ,$$

i.e., if α alone can be considered as the conditioning constant for the BVP (2.1,2)? Under mild scaling assumptions for the BC, this is indeed true, as discussed in [7] (which contains a more careful teatment of several concepts introduced in this section).

Since well-conditioned BVPs have a dichotomy, it is possible to find a fundamental solution matrix $\Phi = [\Phi^- \Phi^+] := YC$ such that the $n \times (n-r)$ matrix Φ^- involves decreasing components and the $n \times r$ matrix Φ^+ involves r increasing components. A fundamental solution matrix $\tilde{Y} = [Y^1 Y^2]$, partitioned as is Φ , is called *consistent* if $span\ Y^1(0) \cap span\ \Phi^+(0) = 0$ (where $span\ C =$ the vector space spanned by the columns of C), so $Y^1(x)$ contains only decreasing components if the BVP is well-conditioned. Finally, in this case one can argue (e.g., see [1,8]) that if a homogeneous solution $B(x)$ satisfies $B_0\phi(0) = 0$, then $\phi \notin span\ \Phi^-$, i.e., $\phi(x)$ is not a decreasing solution component.

III. "One-Step" Discretisations

Recall that the simple shooting method for solving (2.1,2) involves solving IVPs for a fundamental solution matrix $Y(x)$ and particular solution matrix $\tilde{y}(x)$, viz. for some nonsingular matrix E and vector v solve

$$\mathbf{L}Y = 0 \qquad 0 < x < 1 \ , \ Y(0) = E$$

$$\mathbf{L}\tilde{y} = f \qquad 0 < x < 1 \ , \ \tilde{y}(0) = v \ ,$$

and then find

$$y = Y(\mathbf{B}Y)^{-1}(b - B\tilde{y}) + \tilde{y} \ .$$

Computational difficulty frequently arises, of course, because the increasing components of Y dominate.

The multiple shooting method [2,11] attempts to circumvent this difficulty by requiring that IVPs only be solved over smaller subintervals. In particular, given

$$0 = x_1 < x_2 \ , \ \cdots < x_N < x_{N+1} = 1 \ , \tag{3.1}$$

one solves for a fundamental solution matrix Y_j and particular solution y_j from the

IVPs

$$\mathbf{L}Y_j = 0 \quad x_j < x < x_{j+1} \ , \ Y_j(x_j) = E_j \ ,$$

$$\mathbf{L}y_j = f \quad x_j < x < x_{j+1} \ , \ y_j(x_j) = v_j \ , \tag{3.2}$$

and then

$$y(x) = Y_j(x)\alpha_j + y_j(x) \quad x_j < x < x_{j+1} \quad (1 \le j \le N) .$$

The coefficients $\alpha_1, \ldots, \alpha_N$ are determined by requiring that $y(x)$ be continuous at $x_2, \ldots, x_N$ and satisfy the BC (2.2). Defining (for later convenience) E_{N+1} such that $y(x_{N+1}) = E_{N+1}\alpha_{N+1}$, these conditions can be expressed as a system of equations for $\alpha = (\alpha_1, \ldots, \alpha_{N+1})^T$ of the form

$$M\alpha = d \tag{3.3}$$

where

$$M = \begin{bmatrix} B_0E_1 & & & & B_1E_{N+1} \\ -F_1 & E_2 & & & \\ & -F_2 & E_3 & & \\ & & \ddots & & \\ & & & -F_N & E_{N+1} \end{bmatrix} , \tag{3.4}$$

$$F_i = Y_i(x_{i+1}) \quad (1 \le i \le N) .$$

It is fairly direct to show the following result [8,9].

Theorem 1. For the "standard" multiple shooting matrix $\overline{M}$ (each $E_i = I$, $v_i = 0$ in (3.2)),

$$\overline{M}^{-1} = \begin{bmatrix} Y_1(x_1)(\mathbf{B}Y_1)^{-1} & G(x_1,x_2) & \cdots & G(x_1,x_{N+1}) \\ Y_1(x_2)(\mathbf{B}Y_1)^{-1} & G(x_2,x_2) & \cdots & G(x_2,x_{N+1}) \\ \vdots & \vdots & & \vdots \\ Y_1(x_{N+1})(\mathbf{B}Y)^{-1} & G(x_{N+1},x_2) & \cdots & G(x_{N+1},x_{N+1}) \end{bmatrix} , \tag{3.5}$$

where $G(x,t)$ is defined in (2.6). Thus, if $||Y_i(x_{i+1})|| \leq k \; (1 \leq i \leq N)$, then

$$cond(\overline{M}) = ||\overline{M}||_\infty ||\overline{M}^{-1}||_\infty \leq (k+1)(\beta+N\alpha) , \tag{3.6}$$

with β , α defined in (2.7,8).

In other words, if the mesh (3.1) is sufficiently fine that k is of reasonable size, then $cond(\overline{M}) = O(N)$. This is as small as can be generally expected, and the conditioning of the discretization matrix $\overline{M}$ reflects the conditioning of the underlying BVP itself. Similar remarks hold for the general multiple shooting matrix (3.4) since

$$M^{-1} = \begin{bmatrix} E_1^{-1} & & & \\ & E_2^{-1} & & \\ & & \ddots & \\ & & & E_{N+1}^{-1} \end{bmatrix} \overline{M}^{-1} . \tag{3.7}$$

The constant k reflects the sensitivity of the IVPs in (3.2). For example, one can show that $k \leq e^{ch}$ where $c = \max_{0 \leq x \leq 1} ||A(x)||_\infty$ and $h = \max_{1 \leq i \leq N} (x_{i+1}-x_i)$, so that if $h \leq \frac{\Delta}{N}$ as $N \to \infty$ then $k = 1 + \frac{c\Delta}{N} + O(N^{-2})$. This restriction that $h||A||$ is of reasonable size is an unrealistic assumption for a large class of problems, however (see section VII).

The above results extend to the case of one-step finite difference methods, since they give matrices $\tilde{M}$ with the same structure as M (for the mesh (3.1). For the BVP (2.1,2), define *stability* as the property that for $h = \max_{1 \leq i \leq N} (x_{i+1}-x_i)$ sufficiently small, the discretization matrices $\tilde{M}$ satisfy

$$||\tilde{M}^{-1}||_\infty \leq \kappa(\beta,\alpha)$$

where $\kappa(\beta,\alpha)$ is an algebraic expression in β and α (not involving "much larger" quantities like $e^{||A||}$ or e^{α}). Thus, the conditioning of a stable discrete problem reflects that of the BVP. From Theorem 1, it follows [13] that

Corollary. Any consistent one-step discretization is stable.

This is the same stability result as for IVPs [14], but we use this slightly stronger definition of stability (see also [15]). A by-product of the proof of this corollary is that $\tilde{M}^{-1}$, the so-called discrete Green's function, is explicitly constructed. This is useful, e.g., in comparing the truncation error τ to the actual error e (since $Me = \tau$). Finite difference methods for which $||e||$ is higher order than $||\tau||$ are considered in [16]. Construction of the discrete Green's function for a second order BVP results in interesting observations being made about mesh selection in [17] and about an iteration strategy for the nonlinear problem in [12].

The (spline) collocation method for solving (2.1,2) involves determining a spline function satisfying (2.1) at a discrete set of points and satisfying (2.2). For example, require $s(x) \in C[0,1]$ to be of order $k+1$ (degree $\leq k$) on each subinterval of (3.1) and determine $s(x)$ from the conditions $\mathbf{B}s = b$,

$$\mathbf{L}s(x_{ij}) = f(x_{ij}) \text{ , } where \ x_i \leq x_{i1} < \cdots < x_{ik} \leq x_{i+1} \ (1 \leq i \leq N) \ . \quad (3.8)$$

Collocation is equivalent to an implicit Runge-Kutta method (e.g, [18]), and there are several ways to view such a method, each with its own theoretical or computational advantages. One is the following: Letting $y_i := s(x_i)$ and $y_{ij} := s(x_{ij})$, the collocation conditions (3.8) and continuity conditions can be expressed as

$$y_{i+1} = y_i + h_i \sum_{l=1}^{k} \beta_l (A(x_{il}) y_{il} + f(x_{il}))$$

$$y_{ij} = y_i + h_i \sum_{l=1}^{k} \alpha_{jl} (A(x_{il}) y_{il} + f(x_{il})) \tag{3.9}$$

where $h_i := x_{i+1} - x_i$ and $\{\beta_l, \alpha_{jl}\}$ are known constants. Since the BVP is linear, (3.9) can be condensed to the form $y_{i+1} = F_i y_i + g_i$ for an appropriately defined $n \times n$ matrix F_i and vector g_i. Hence, collocation may be viewed as a one-step discretization method, and the above techniques can be used to construct the explicit inverse for the collocation matrix.

These stability results for well-conditioned BVPs have been extended for higher order differential equations for finite differences [19] and collocation [20,13]. Still, further analysis using the explicit discrete Green's function should provide many additional insights, both for "smooth" problems (where $h||A(x)||$ is assumed to be small throughout $[0,1]$) and for singular perturbation problems.

IV. Marching Algorithms

Here we consider a class of methods which, as with multiple shooting, involve solving IVPs over subintervals. The selection of the subintervals (or mesh (3.1)) is done as the integration is proceeding through the interval.

For the main class of marching algorithms which we consider, the jth step is as follows: Given $\tilde{E}_j$ and $\tilde{v}_j$, solve

$$\mathbf{L}\tilde{Y}_j = 0 \quad x_j < x < x_{j+1}, \ \tilde{Y}_j(x_j) = \tilde{E}_j \tag{4.1}$$

$$\mathbf{L}\tilde{y}_j = f \quad x_j < x < x_{j+1}, \ \tilde{y}_j(x_j) = \tilde{v}_j .$$

Usually, $\tilde{Y}_j$ is a subset of the full fundamental solution matrix and the stability of the computation is closely tied to the solution dichotomy. Letting $y = Y\alpha_j + \tilde{y}_j$ on $[x_j, x_{j+1}]$, the coefficients $\alpha_1, \ldots, \alpha_N$ are determined in a similar way as for multiple shooting (which can in itself be viewed as a marching algorithm). Important features of marching are that i) the number of IVPs to be solved and dimension of the resulting matrix system are normally less than for multiple shooting and ii) the mesh points (3.1) can be chosen adaptively to prevent undue loss of accuracy.

The most popular marching algorithm has been the Godunov-Conte algorithm or stabilized march [8,21]. We describe a version of this algorithm for the case of separated BC

$$B_0 = \begin{bmatrix} B_0^1 \\ 0 \end{bmatrix}, B_1 = \begin{bmatrix} 0 \\ B_1^2 \end{bmatrix}, b = \begin{bmatrix} b_1 \\ b_2 \end{bmatrix}, \tag{4.2}$$

where B_0^1 is $k \times n$, B_1^2 is $(n-k) \times n$, and b_1 is a k vector. First, one computes the $n \times k$ so-called complementary solution matrix $\tilde{Y}_1$ (with linearly independent columns) and particular solution vector $\tilde{y}_1$, satisfying

$$\mathbf{L}\tilde{Y}_1 = 0 \quad x_1 < x < x_2 \ , \ B_0\tilde{Y}_1(0) = 0$$

$$\mathbf{L}\tilde{y}_1 = f \quad x_1 < x < x_2 \ , \ B_0\tilde{y}_1(0) = b_1 \ .$$

At x_2 , a factorization of $\tilde{Y}_1(x_2)$ of the form

$$\tilde{Y}_1(x_2) = [\tilde{E}_2' \ \tilde{E}_2] \begin{bmatrix} 0 \\ G_2 \end{bmatrix} \tag{4.3}$$

is computed, * where $[\tilde{E}_2' \ \tilde{E}_2]$ is an $n \times n$ orthogonal matrix, $\tilde{E}_2$ is $n \times k$, and G_2 is $k \times k$. ** A vector $\tilde{v}_2$ orthogonal to $\tilde{E}_2$ is computed. One continues these steps until $x_{s+1} = 1$ is reached.

---

* we write this, instead of, say $\begin{bmatrix} G_2 \\ 0 \end{bmatrix}$, for later convenience.

** Of importance is not so much orthogonality as separation of the subspace.

From section 3, if the BVP is well-conditioned then the condition $B_0\tilde{Y}_0(0) = 0$ implies that $\tilde{Y}_0(x)$ involves no decreasing solution components, i.e., there is an consistent fundamental solution matrix $Y_0(x) = [Y_0^- Y_0^+]$ with $\tilde{Y}_0 = Y_0^+$. From (4.3) this property is preserved on the next subinterval (and subsequent ones). Therefore, the *separated BC effect a decoupling* of fundamental solution components into decreasing and increasing terms.

A different type of marching algorithm which will be related to the stabilized march and multiple shooting in the next section is invariant imbedding. To describe this method for the separated BC case (4.2), rewrite the BVP as

$$\begin{bmatrix} y_1 \\ y_2 \end{bmatrix}' - \begin{bmatrix} A_0 A_1 \\ A_2 A_3 \end{bmatrix} \begin{bmatrix} y_1 \\ y_2 \end{bmatrix} = \begin{bmatrix} f_1 \\ f_2 \end{bmatrix} \quad 0 < x < 1 , \tag{4.4}$$

$$[B_{01} \;\; B_{02}] \begin{bmatrix} y_1(0) \\ y_2(0) \end{bmatrix} = b_1 \quad [B_{11} \;\; B_{12}] \begin{bmatrix} y_1(1) \\ y_2(1) \end{bmatrix} = b_2 , \tag{4.5}$$

where B_{01} , $A_0(x)$ are $k \times k$ matrices and B_{01} is nonsingular, B_{12} , $A_3(x)$ are $(n-k)\times(n-k)$ matrices, y_1 , $f_1(x)$ are k-vectors, and the partitioning is otherwise clear.

If R , v , and y_2 satisfy the IVPs

$$R' = [A_0 - RA_2]R - RA_3 + A_2 \tag{4.6}$$

$$B_{01}R(0) = -B_{02} , \tag{4.7}$$

$$v' = [A_0 - RA_2]v - Rf_2 + f_1 \tag{4.8}$$

$$B_{01}v(0) = b_1 , \tag{4.9}$$

and

$$y_2' = [A_3+A_2R]y_2 + A_2v + f_2 \tag{4.10}$$

$$[B_{11}R(1) + B_{12}]y_2(1) = b_2-B_{12}v(1) \,, \tag{4.11}$$

then one can verify directly that $y = \begin{bmatrix} y_1 \\ y_2 \end{bmatrix}$ satisfies (4.4,5) if

$$y_1 = Ry_2 + v \,. \tag{4.12}$$

Moreover, if $Y = \begin{bmatrix} Y_0^- & Y_1^+ \\ Y_2^- & Y_3^+ \end{bmatrix}$ is a consistent fundamental solution matrix satisfying $[B_{01} \;\; B_{02}] \begin{bmatrix} Y_1^+(0) \\ Y_3^+(0) \end{bmatrix} = 0$, $Y_3^+(0)$ nonsingular, then the *Riccati* matrix $R(x)$ (satisfying the Riccati equation (4.6)) is

$$R = Y_1^+(Y_3^+)^{-1} \quad 0 < x < 1 \,. \tag{4.13}$$

In practice, this method must be implemented as a marching procedure because there is no assurance that the solution to this nonlinear Riccati equation remains bounded (or, in other words, that $Y_3^+(x)$ remains nonsingular). When $R(x)$ becomes large, one can perform a change of imbedding (or change of variables) and continue the integration of (4.6) until trouble arises again. Observe that, choosing subintervals in this way such that R remains bounded, the linear IVPs (4.8) and (4.10) can be integrated (see [22] for further discussion and references).

One way to derive the invariant imbedding equations is to attempt to decouple decreasing and increasing components of (4.4) by transforming the coefficient matrix to block upper triangular form. The key equation involving this transformation is the *Lyapunov equation*, whose solution is the Riccati matrix (e.g., see [1]).

V. Equivalences for a Particular Factorization

Now we show that the methods presented in the previous two sections are closely connected with each other. Consider the standard multiple shooting method applied to (4.4,5), and with the obvious partitioning let $Y_i(x_{i+1}) =: \begin{bmatrix} Y_i^0 & Y_i^1 \\ Y_i^2 & Y_i^3 \end{bmatrix}$. Before solving the resulting linear system by Gaussian elimination, a natural reordering giving a banded matrix with "large" diagonal elements is

$$M' = \begin{bmatrix} B_{01} & B_{02} & & & & \\ -Y_1^2 & -Y_1^3 & 0 & I & & \\ -Y_1^0 & -Y_1^1 & I & 0 & & \\ & & -Y_2^2 & -Y_2^3 & 0 & I \\ & & & & & \ddots \end{bmatrix}. \tag{5.1}$$

For the first elimination step, perform the row and column elimination $[B_{01} \;\; B_{02}] \rightarrow [I \;\; -R_1]$ on the first block (so $R_1 = -B_{01}^{-1}B_{02}$). Then (*) eliminate the column blocks below the identity block, so that for appropriately defined $\tilde{Y}_1^1$ and $\tilde{Y}_1^3$, M' is of the form

$$\begin{bmatrix} I & -R_1 & & & & \\ 0 & -\tilde{Y}_1^3 & 0 & I & & \\ 0 & -\tilde{Y}_1^1 & I & 0 & & \\ & & -Y_2^2 & -Y_2^3 & 0 & I \\ & & & & & \ddots \end{bmatrix}.$$

Next, using pivoting with row interchanges on the blocks $\begin{bmatrix} -\tilde{Y}_1^3 \\ -\tilde{Y}_1^1 \end{bmatrix}$, eliminate the bottom block. For any row interchanges done in this step, perform corresponding column interchanges in the third and fourth column blocks, resulting in

$$\begin{bmatrix} I & -R_1 & & & & \\ 0 & -\hat{Y}_1^3 & 0 & I & & \\ 0 & 0 & I & -\hat{R}_2 & & \\ & & -Y_2^2 & -Y_2^3 & 0 & I \\ & & & & & \ddots \end{bmatrix}. \tag{5.2}$$

One now continues as from (*) to complete the block upper triangularization of M' .

It is straightforward to see [8] that this factorization with alternate row and column interchanges on blocks "almost" chooses the largest element as a pivot (viz. it uses Markowich pivots) and can thus be expected to be stable from a result of van Veldhuizen [23]. Moreover, the following equivalences hold [8]: In the absence of errors, if no pivoting is necessary then after factorization the upper right blocks R_i of the 2×2 diagonal blocks are equal to the Riccati matrix values $R(x_i)$ from invariant imbedding, and the blocks $\begin{bmatrix}\tilde{Y}_i^1\\ \tilde{Y}_i^3\end{bmatrix}$ are equal to the complementary solution $\tilde{Y}(x_i)$ from the stabilized march (with $\tilde{E}_2 = \begin{bmatrix}R_2\\ I\end{bmatrix}$, $G_2 = \tilde{Y}_2^3$ in (4.3)). Pivoting plays the role of a change of variables, e.g., the change of variables corresponding to (5.2) gives $\hat{R}_2 = \hat{R}(x_2)$ from invariant imbedding, $\hat{Y}(x_2) = \begin{bmatrix}\hat{Y}_1^1\\ \hat{Y}_1^3\end{bmatrix}$ from stabilized march. Other more specialized connections of invariant imbedding to other methods have been shown before, e.g., [22,24]. Note that the results still hold asymptotically if M' is only a multiple shooting-like matrix (arising, e.g., from a one-step finite difference discretization).

VI. A General Factorization and Decoupling

In the matrix factorization of the previous section, the separated BC were implicitly used to decouple the growing and decreasing (fundamental solution) components. For general BC, this is not possible. Fortunately, the standard matrix factorization algorithms can themselves effect this same decoupling.

To see how this decoupling occurs, consider the case of a 2×2 matrix

$$A = [v_1 \;\; v_2] = [w_1 \;\; w_2] \begin{bmatrix}\alpha_1 & \beta_1\\ \alpha_2 & \beta_2\end{bmatrix} ,$$

where v_1 , v_2 are linearly independent vectors, α_1 , α_2 , β_1 , β_2 are constants such that $|\alpha_1|^2 + |\alpha_2|^2$ and $|\beta_1|^2 + |\beta_2|^2$ are $O(1)$, and w_1 , w_2 are vectors such that $||w_1|| >> ||w_2||$. Note that v_1 and v_2 point in roughly the same direction (or "make a small angle between them"). If A is factored into

$$A = [u_1 \quad u_2] \begin{bmatrix} \gamma_1 & \gamma_2 \\ 0 & \gamma_3 \end{bmatrix} \tag{6.1}$$

where the vectors u_1 , u_2 do not make a small angle between them and $||u_1|| \approx ||u_2||$ then geometrical arguments give that $\gamma_1 \approx ||w_1||$, $\gamma_3 \approx ||w_2||$. This extends to general block factorizations [1] and is the basis for the particular decoupling algorithm of Mattheij which we outline below.

Consider the linear BVP (2.1,2) with

$$B_0 = \begin{bmatrix} B_{01} & B_{02} \\ B_{03} & B_{04} \end{bmatrix} , B_1 = \begin{bmatrix} B_{11} & B_{12} \\ B_{13} & B_{14} \end{bmatrix} , \tag{6.2}$$

where the same partitioning for B_0 and B_1 is done (in a manner discussed below). Using standard multiple shooting (modification to more general methods is straightforward) yields the matrix

$$M = \begin{bmatrix} B_{01} & B_{02} & & & & B_{11} & B_{12} \\ & F_1 & -I & & & & \\ & & F_2 & -I & & & \\ & & & \ddots & F_N & & -I \\ B_{03} & B_{04} & & & & B_{13} & B_{14} \end{bmatrix} . \tag{6.3}$$

Choosing orthogonal Q_1 by appropriately factoring B_0 (see [1] for details), now find orthogonal $Q_2, \ldots, Q_{N+1}$ by successively computing block QR factorizations of $F_i Q_i$, viz.

$$A_i Q_i = Q_{i+1} \Phi_i \qquad i = 1, \ldots, N , \tag{6.4}$$

((6.4) may be interpreted as a discrete approximation to the Lyapunov equation) where

$\Phi_i = \begin{bmatrix} \Phi_i^+ & \Phi_i^\pm \\ 0 & \Phi_i^- \end{bmatrix}$ are partitioned as in (6.2) and $||\Phi_i^+||$ and $||\Phi_i^-||$ correspond to the magnitudes of growing and decaying components, respectively (and are monitored in the first place to determine the correct partitioning in (6.2) [25]). Now, as compared to (5.2), (6.3) has become

$$\begin{bmatrix} 0 & \tilde{B}_{02} & & & & & \tilde{B}_{11} & \tilde{B}_{12} \\ \Phi_1^+ & \Phi_1^\pm & -I & 0 & & & & \\ 0 & \Phi_1^- & 0 & -I & & & & \\ & & & & \ddots & & & \\ & & & & \Phi_N^+ & \Phi_N^\pm & -I & 0 \\ & & & & 0 & \Phi_N^- & 0 & -I \\ \tilde{B}_{03} & \tilde{B}_{04} & & & & & \tilde{B}_{13} & \tilde{B}_{14} \end{bmatrix} . \tag{6.5}$$

It is now easy to obtain an LU factorization of M , so the multiple shooting system can be solved directly. An important feature of this factorization method is that the decoupling A provides a *stable* way to use a compactification or condensing algorithm [1,26] which is critical for large problems where storage requirements for "standard" factorizations are exorbitant. This factorization approach also provides motivation for another marching type multiple shooting method [25] for which an efficient code has been written [27].

VII. Conclusion

Well-conditioning of linear BVPs for ODEs has been discussed, and we have pointed out that well-conditioned problems must have a dichotomy. This result will undoubtedly play a fundamental role in future developments, e.g., in a characterization of what types of turning point problems [28] are well-conditioned. The standard numerical methods for solving smooth BVPs - multiple shooting, stabilized march, collocation, finite differences, and invariant imbedding - have been seen to be closely related, and under certain (often quite restrictive) conditions their matrices inherit the well-conditioning of the BVPs. Understanding of these close relationships, which only occurred fairly recently, followed from studying the complete algorithms for solving BVPs, viz. the discretization method *and* the solution method for the resulting linear system of equations.

Normally, one role of the linear system factorization is to decouple the increasing and decreasing components. This is automatically done for well-conditioned BVPs if the BC are separated. However, when the BC are only partially separated, a factorization which preserves the nonzero structure of the matrix at the expense of choosing large pivot elements should be used with caution, since it may only partially decouple solution components (e.g., see [1]) and hence be unstable.

Although consideration has basically been restricted to smooth problems (with $h||A||$ small) in the previous section, interesting tools have been provided for analyzing numerical methods for singular perturbation problems [28]. Realizing that a successful method must appropriately preserve the underlying dichotomy in the discrete system and must utilize decoupling for stability, we see that the discrete Green's function can provide a general framework from which to analyze most methods.

Successful methods for singular perturbation problems perform the decoupling (i) of the slow versus fast components or (ii) of the fast decreasing and fast increasing components. This can of course be done in a variety of ways. In [30], a decoupling (i) is done a priori via an analytical coordinate transformation, whereby the Riccati matrix solving a resulting Lyapunov equation is calculated. In [28], $A(x)$ is transformed throughout $[0,1]$, and decoupling information from its eigenvalues used to choose different one-sided or symmetric difference schemes for the various components. In [18] (and references therein), symmetric collocation schemes are used on meshes chosen to assure that the solution components (satisfying a dichotomy) are sufficiently well approximated on the appropriate subintervals. Future study of the many different combinations of discretizations and decouplings (e.g., see also [29,31]) in order to determine

which are the most efficient or reliable for the various types of singular perturbation problems will be most interesting.

Acknowledgement: **The author is grateful to L.F. Shampine for his, as usual, helpful comments.**

References

1. R.M.M. Mattheij, Decoupling and stability of BVP algorithms, to appear in SIAM Review.

2. H.B. Keller, Numerical Methods for Two Point Boundary Value Problems, Blaisdell, New York, 1968.

3. A.K. Aziz (ed.), Numerical Solutions of Boundary Value Problems for Ordinary Differential Equations, Academic Press, New York, 1975.

4. B. Childs et al. (ed.), Codes for Boundary-Value Problems in Ordinary Differential Equations, Lecture Notes in Computer Science 76, Springer, Berlin, 1978.

5. H.B. Keller, Numerical Solution of Two-point Boundary Value Problems, SIAM Regional Conference Series 24, Philadelphia, 1976.

6. SIAM Journal of Numerical Analysis, Vol. 14, No. 1 (1977). (Special issue on BVODEs.)

7. F. de Hoog and R.M.M. Mattheij, Conditioning, dichotomy, and scaling for two-point BVPs, these proceedings.

8. M. Lentini, M.R. Osborne, and R.D. Russell, The close relationships between methods for solving two-point boundary value problems, SIAM J. Numer. Anal. (1984).

9. R.M.M. Mattheij, Estimates for the errors in the solution of linear boundary value problems due to perturbations, Computing 27 (1981), 229-318.

10. F.R. de Hoog and R.M.M. Mattheij, On dichotomy and well-conditioning in BVP, Report 8356, Kath. Univ., Nijmegen (1983).

11. M.R. Osborne, On shooting methods for boundary value problems, J. Math. Anal. Appl. 27 (1969), 417-433.

12. R.M.M. Mattheij and M.D. Smooke, Estimates for the inverse of tridiagonal matrices arising in boundary value problems, to appear.

13. J. Paine and R.D. Russell, Conditioning of collocation matrices and discrete Green's functions, to appear.

14. P. Henrici, Discrete Variable Methods in Ordinary Differential Equations, John Wiley, New York, 1962.

15. C. de Boor, F. de Hoog, and H.B. Keller, The stability of one-step schemes for first-order two-point boundary value problems, SIAM J. Numer. Anal. 20 (1983), 1139-1146.

16. H.-O. Kreiss, T.A. Manteuffel, B. Swartz, B. Wendroff, and A.B. White, Jr., Supraconvergent schemes on irregular grids, Los Alamos Report LA-UR-83-2818, 1983.

17. F. de Hoog and D. Jackett, On the rate of convergence of finite difference schemes on non-uniform grids, J. Austral. Math. Soc. Series B, to appear.

18. U. Ascher, Two families of symmetric difference schemes for singular perturbation problems, these proceedings.

19. C. de Boor and F. de Hoog, Stability of numerical schemes for two-point boundary value problems, to appear.

20. U. Ascher, S. Pruess, and R.D. Russell, On spline basis selection for solving differential equations, SIAM J. Numer. Anal. 20 (1983), 121-142.

21. M.R. Scott and H.A. Watts, Computational solution of linear two point boundary value problems via orthonormalization, SIAM J. Numer. Anal. 14 (1977), 40-70.

22. H.B. Keller and M. Lentini, Invariant imbedding, the Box scheme and an equivalence between them, SIAM J. Numer. Anal. 19 (1982), 942-962.

23. M. van Veldhuizen, A note on partial pivoting and Gaussian elimination, Numer. Math. 29 (1977), 1-10.

24. I. Babuška, The connection between the finite difference like methods and the methods based on initial value problems for O.D.E., in [3].

25. R.M.M. Mattheij and G.W.M. Staarink, An efficient algorithm for solving general linear two point BVP, SIAM J. Sci. Stat. Comp. 5 (1984).

26. P. Deuflhard and G. Bader, Multiple-shooting techniques revisited, in Numerical Treatment of Inverse Problems in Differential and Integral Equations, P. Deuflhard and E. Hairer (eds.), Birkhauser, Boston, 1983.

27. G.W.M. Staarink and R.M.M. Mattheij, BOUNDPACK: A package for solving boundary value problems, Math. Inst., Kath. Univ, Nijmegen, 1982.

28. H.-O. Kreiss, N.K. Nichols, and D.L. Brown, Numerical methods for stiff two-point boundary value problems, to appear.

29. R.E. O'Malley, Jr., On the simultaneous use of asymptotic and numerical methods to solve two-point problems with boundary and interior layers, these proceedings.

30. R.M.M. Mattheij and R.E. O'Malley, Jr., On solving boundary value problems for multi-scale systems using asymptotic approximations and multiple shooting, to appear in BIT (1984).

31. V. Majer, Numerical solution for boundary value problems for ordinary differential equations of nonlinear elasticity, Ph.D. Thesis, Univ. Maryland, 1984.

Progress in Scientific Computing, Vol. 5
Numerical Boundary Value ODEs

The Role of Conditioning in Shooting Techniques

Frank de Hoog and Robert Mattheij

Abstract. This paper examines shooting and multiple shooting as a technique for the analysis of numerical schemes applied to two point boundary value problems. The aim of the analysis is to deduce convergence of a numerical scheme by establishing the convergence of the scheme when applied to a number of subproblems. Since such an approach is useful only if the subproblems are reasonably well conditioned, the question of conditioning is addressed. It is shown that there exist subproblems that are at least as well conditioned as the original problem. Two examples are presented for which the shooting approach leads to a substantial simplification in the analysis.

1. Introduction

We shall consider the system of ordinary differential equations

$$(1.1)\quad L\,y := \dot{y} - Ay = f \;,\; 0 < t < 1$$

subject to the boundary conditions

$$(1.2)\quad B\,y := B_0 y(0) + B_1 y(1) = b$$

where for some $p \in [1,\infty]$, $A \in [L_p(0,1)]^{n\times n}$ and $f \in [L_p(0,1)]^n$ while $B_0, B_1 \in \mathbb{R}^{n\times n}$ and $b \in \mathbb{R}^n$.

Any solution to (1.1) has the form (see for example Keller [4])

$$y = Yc + \hat{y}$$

where $c \in \mathbb{R}^n$ and $Y \in [L_p^1(0,1)]^{n\times n}$, $\hat{y} \in [L_p^1(0,1)]^n$ are the solutions of the initial value problems

$$LY = 0 \quad , \quad Y(0) = I$$

$$L\hat{y} = f \quad , \quad \hat{y}(0) = 0$$

On substituting into the boundary condition (1.2) we obtain the equation

$$(\ B\ Y)c = b - B\ \hat{y}$$

and it follows that the boundary value problem (1.1), (1.2) has a unique solution if and only if the matrix $B\ Y$ is nonsingular. In the sequel we shall assume that this is the case. Then, on writing

$$\Phi := Y(\ B\ Y)^{-1}$$

and noting that

$$\hat{y} = \int_0^t Y(t)\ Y^{-1}(s)\ f(s)\ ds = \int_0^t \Phi(t)\ \Phi^{-1}(s)\ f(s)\ ds$$

we find that

$$y(t) = \Phi(t)b + \hat{y}(t) - \Phi(t)\ B\ \hat{y}$$

which can be rewritten as

$$(1.3) \quad y(t) = \Phi(t)b + \int_0^1 G(t,s)\ f(s)\ ds$$

where

$$G(t,s) = \begin{cases} \Phi(t)\ B_0\ \Phi(0)\ \Phi^{-1}(s) & , \quad t > s \\ -\Phi(t)\ B_1\ \Phi(1)\ \Phi^{-1}(s) & , \quad t < s \end{cases}$$

or equivalently

$$G(t,s) = \begin{cases} Y(t)\ E\ Y^{-1}(s) & , \quad t > s \\ Y(t)\ (E-I)\ Y^{-1}(s) & , \quad t < s \end{cases}$$

with $E = \Phi(0)\, B_0 = (\, B\, Y)^{-1}\, B_0$. Thus, in principle, a knowledge of the fundamental solution allows us to write down the solution of the boundary value problem (1.1), (1.2).

A knowledge of how perturbations in the right hand sides of (1.1) and (1.2) affect the solution y is obviously important when solving (1.1), (1.2) numerically. In order to characterise this mathematically, we introduce the following norms. Let $|.|$ denote the usual Euclidean norm of a vector and for $f \in [L_p(0,1)]^n$ define

$$\|f\|_p = \{\int_0^1 |f(s)|^p \, ds\}^{1/p}$$

with its limiting value

$$\|f\|_\infty = \sup_t |f(t)| \, .$$

It now follows from (1.3) that

$$(1.4) \quad \|y\|_\infty \leq \beta\, |b| + \alpha_q\, \|f\|_p \, , \quad \frac{1}{p} + \frac{1}{q} = 1$$

where

$$(1.5) \quad \beta = \|\Phi\|_\infty$$

and

$$(1.6) \quad \alpha_q = \sup_t \left\{ \int_0^1 \left| G(t,s) \right|^q ds \right\}^{1/q}.$$

The constants β and α_q in equation (1.4) give a bound on how perturbations in the right hand sides of (1.1) and (1.2) may be amplified and thus play the role of condition numbers for the problem. The choice of p depends to a large extent on the problem in question. For example, if

$$L\, y = y' + y/\varepsilon \quad , \quad B\, y = y(0)$$

then

$$\beta = 1 \;, \quad \alpha_\infty = 1 \;, \quad \alpha_1 = \varepsilon\left(1 - e^{-1/\varepsilon}\right)$$

and it is usually most convenient to take $p = \infty$ if ε is small. On the other hand, if $\varepsilon = O(1)$ then the choice $p = 1$ may be more appropriate. In some applications, neither choice is appropriate. For example if an mth order equation is written as a first order system we may be interested only in perturbations of the solution to the mth order equation. In this case we could deduce a stability result from (1.3) using 'weighted norms'. Specificaly, we could use

$$\| W_1 y \|_\infty \leq \beta \; \left| W_2 b \right| \; + \; \alpha_q \; \| W_3\, f \|_p$$

where

$$\beta = \| W_1 \Phi W_2^{-1} \|_\infty$$

and

$$\alpha_q = \sup_t \{ \int_o^1 | W_1 \ G(t,s) \ W_3^{-1} |^q \ ds \}^{1/q} \ .$$

However for simplicity we consider estimates of the form (1.4) in the present paper.

The aim of this paper is to examine the role played by the condition numbers β and α_∞ in the analysis of numerical schemes by shooting techniques. Since the analysis draws fairly heavily on some recent work by de Hoog and Mattheij [3] on the connection between dichotomy and conditioning, we review some of their results in sections 2 and 3. In particular, we examine scaling of the boundary condition (1.2) in section 2 while in section 3 we derive the result in [3] that for every well conditioned problem (1.1), (1.2) there exist separated boundary conditions such that the resulting boundary value problem is still reasonably well conditioned. In section 4 we present an analysis of shooting and multiple shooting that explicitly shows the dependence on the condition numbers β and α_∞. Finally, some closing remarks are made in section 5.

2. Scaling of Boundary Conditions

While the scaling of the differential operator $\mathbb{L}$ defined by (1.1) is unambiguous in that the coefficient of the term y' is the identity, the scaling of the boundary conditions (1.2) is somewhat less satisfactory. The usual practice is to impose a normalisation such as

$$(2.1)\qquad \max\{\,|B_0|\,,\,|B_1|\,\} = 1$$

but this is not entirely satisfactory as is demonstrated by the example

$$B_0 = \begin{bmatrix} 1 & 0 \\ 0 & 10^{-2} \end{bmatrix}, \qquad B_1 = \begin{bmatrix} 1 & 0 \\ 0 & 10^{-3} \end{bmatrix}$$

It is clear that a more appropriate scaling for these boundary conditions is

$$B_0 = \begin{bmatrix} 1 & 0 \\ 0 & 1 \end{bmatrix}, \qquad B_1 = \begin{bmatrix} 1 & 0 \\ 0 & 10^{-1} \end{bmatrix} .$$

If both B_0 and B_1 are diagonal an intuitively appealing way to scale the boundary conditions is to require that the maximum element in modulus of each row of $[B_0|B_1]$ is one. Although it is generally not possible to scale B_0 and B_1 such that they are diagonal, it turns out that it is possible to obtain a scaling such that B_0 and B_1 retain the essential character of a diagonal matrix. Specifically we have

Lemma 2.1 (de Hoog and Mattheij [3]). **Let B_0, $B_1 \in R^{n\times n}$ and rank $[B_0|B_1] = n$. Then, there exists a nonsingular matrix S such that**

$$B_0 = S\, Z_0\, Q_0^T \quad , \quad B_1 = S\, Z_1\, Q_1^T$$

where Q_0, Q_1 are orthogonal matrices and Z_0, Z_1 are diagonal matrices with the structure

$$\text{(2.2a)} \qquad Z_0 = \left[\begin{array}{c|c} Z_{00} & 0 \\ \hline 0 & I_k \end{array}\right] \quad , \quad Z_1 = \left[\begin{array}{c|c} I_{n-k} & 0 \\ \hline 0 & Z_{11} \end{array}\right]$$

and

$$\text{(2.2b)} \qquad 0 < Z_{00} < 1 \ , \quad 0 < Z_{11} < 1 \ .$$

Thus, without loss of generality we may assume that

$$\text{(2.3)} \qquad B_0 = Z_0\, Q_0^T \ , \quad B_1 = Z_1\, Q_1^T \ .$$

where Q_0, Q_1 are orthogonal and Z_0, Z_1 are diagonal with structure as in (2.2a,b). In the sequel we assume that the boundary conditions have been scaled in this manner. Note that this immediately implies (2.1).

<u>Remark</u> In practice it is more convenient to scale B_0 and B_1 such that the matrix $[B_0|B_1]$ is orthogonal. It then follows that $B_0 = U\, Z_0\, Q_0^T$, $B_1 = U\, Z_1\, Q_1^T$ where U, Q_0, Q_1

are orthogonal matrices and Z_0, Z_1 are diagonal matrices satisfying $Z_0^2 + Z_1^2 = I$. Clearly, this scaling is closely related to that of (2.3) and we have chosen the latter to simplify some of the subsequent algebra. However all of the analysis can easily be modified if the alternative scaling is used.

Now that we have removed the ambiguity in the boundary condition (1.2), we can obtain a bound for β defined by (1.5) in terms of α_∞ defined by (1.6).

Lemma 2.2. Let B_0 and B_1 be scaled as in (2.3). Then

$$\beta = \|\Phi\|_\infty < 2,\alpha_\infty = 2 \sup_{t,s} \left|G(t,s)\right| .$$

Proof. From (2.2a,b),

$$\begin{aligned} \left|\Phi(t)\right| &< \left|\Phi(t)\ (Z_0 + Z_1)\right| \\ &< \left|\Phi(t)\ Z_0\right| + \left|\Phi(t) Z_1\right| \\ &= \left|\Phi(t)\ B_0\right| + \left|\Phi(t)\ B_1\right| \\ &= \left|G(t,0)\right| + \left|G(t,1)\right| \\ &< 2\ \alpha_\infty \end{aligned}$$

#

The inequality (1.4) therefore yields

$$\|y\|_\infty \le 2\,\alpha_\infty\,|b| + \alpha_q\,\|f\|_p\,, \quad \frac{1}{p} + \frac{1}{q} = 1$$

$$\le \alpha_\infty\left(2|b| + \|f\|_p\right)$$

which shows that the single parameter α_∞ serves quite well as the condition number for the boundary value problem (1.1), (1.2).

We conclude this section with the following lemma.

Lemma 2.3 Let $E \in \mathbb{R}^{n\times n}$ and for $0 \le a \le t,\ s \le c \le 1$ define

$$(2.4)\qquad G(t,s) = \begin{cases} Y(t)\,E\,Y^{-1}(s) & , \quad t > s \\ Y(t)\,(E-I)\,Y^{-1}(s) & , \quad t < s \end{cases}$$

Then there exist matrices $\hat{B}_0$ and $\hat{B}_1$ (scaled as in (2.3)) such that the boundary value problem

$$Ly = f\ , \quad a < t < c$$

$$\hat{B}y = \tilde{B}_0 y(a) + \tilde{B}_1 y(c) = b$$

has the solution

$$y(t) = \hat{\Phi}(t)b + \int_a^c G(t,s)\, f(s)\, ds$$

where

$$\hat{\Phi} = Y\,(\,\hat{B}\,Y)^{-1}$$

Proof   On applying Lemma 2.1 to the matrices $E\,Y^{-1}(a)$ and $(I-E)\,Y^{-1}(c)$, we find that there is a nonsingular matrix $\hat{S}$, diagonal matrices $\hat{Z}_0$, $\hat{Z}_1$ with structure as in (2.2a,b) and orthogonal matrices $\hat{Q}_0$, $\hat{Q}_1$ such that

$$E\,Y^{-1}(a) = \hat{S}\,\hat{Z}_0\,\hat{Q}_0^{\,T}\;,\qquad (I-E)\,Y^{-1}(c) = \hat{S}\,\hat{Z}_1\,\hat{Q}_1^{\,T}.$$

It is now easy to verify that

$$\hat{B}_0 = \hat{Z}_0\,\hat{Q}_0^{\,T}\quad,\quad B_1 = \hat{Z}_1\hat{Q}_1^{\,T}$$

yield appropriate boundary conditions. #

The above lemma says two things. Firstly, it says that any G of the form (2.4) is a Green's function of the differential operator L. It also says that for every boundary value problem (1.1), (1.2) there is a boundary value problem on every subinterval (a,c) such that the conditioning of the problem on the smaller interval is at least as good as the conditioning of the problem on the larger interval. For example if we have a singular perturbation problem with a turning point at $t = 1/2$, then on each subinterval (0, 1/2) and (1/2, 1) we can find (in principle at least) boundary conditions for the differential equations on (0,1/2) and (1/2, 1) such that these problems (which now have boundary layers rather than internal layers) are as well conditioned as the original problem.

3. Conditioning of Separated Boundary Value Problems

In this section we derive the result of de Hoog and Mattheij [3] that for every differential equation and associated boundary condition, there are separated boundary conditions such that the conditioning of the resulting boundary value problem is not much worse than the original one. We include this section rather than simply quoting the result for completeness and because the derivation is somewhat different to [3].

As a first step we need to examine how the fundamental solution Φ and the Green's function G defined in section 1 must be modified when the boundary conditions change. Specifically, consider the differential operator

$$L\,y = y' - Ay \tag{3.1}$$

and the boundary conditions

(3.2) $\quad B\,y = B_0 y(0) + B_1 y(1)$

(3.3) $\quad \hat{B}\,y = \hat{B}_0 y(0) + \hat{B}_1 y(1)$.

We associate with (3.1), (3.2) the fundamental solution Φ and the Green's function G while with (3.1), (3.3) we denote these by $\hat{\Phi}$ and $\hat{G}$ respectively.

Clearly,

$$\hat{\Phi} = \Phi\,(\,\hat{B}\,\Phi)^{-1}$$

and hence

(3.4) $\quad B\,\hat{\Phi} = (\,B\,\Phi)(\,\hat{B}\,\Phi)^{-1} = (\,\hat{B}\,\Phi)^{-1}.$

Thus,

(3.5) $\quad \hat{\Phi} = \Phi(\,B\,\hat{\Phi})$.

We now derive a corresponding relation between $\hat{G}$ and G. Let

$$(3.6) \quad \hat{y}_p(t) = \int_0^1 \hat{G}(t,s)\, f(s)\, ds$$

and

$$(3.7) \quad y_p(t) = \int_0^1 G(t,s)\, f(s)\, ds$$

On noting that

$$\hat{y}_p = \Phi c + y_p$$

and imposing the boundary condition

$$\hat{B}\, \hat{y}_p = 0$$

we find that

$$\hat{y}_p = y_p - \Phi(\, \hat{B}\, \Phi)^{-1}\, \hat{B}\;\; y_p \, .$$

Then substituting (3.6) and (3.7) in the above equation and equating the integrands yields

$$(3.8) \quad \hat{G}(t,s) = G(t,s) - \Phi(t)\, (\, \hat{B}\, \Phi)^{-1}\, \hat{B}\, G(\cdot,s) \quad .$$

These relations can now be used to establish the following inequalities.

Lemma 3.1 Let β, α_p and $\hat{\beta}$, $\hat{\alpha}_p$ be the stability constants of (3.1), (3.2) and (3.1), (3.3) respectively. Then,

$$(3.9) \quad \hat{\beta} < \beta \left\{ \left| \hat{\Phi}(0) \right| + \left| \hat{\Phi}(1) \right| \right\}$$

$$(3.10) \quad \hat{\beta} < \alpha_p \left\{ \left| \hat{\Phi}(0) \right| + \left| \hat{\Phi}(1) \right| \right\}$$

$$(3.11) \quad \hat{\alpha}_p < \alpha_p + 2\beta\, \alpha_p \left\{ \left| \hat{\Phi}(0) \right| + \left| \hat{\Phi}(1) \right| \right\}$$

$$(3.12) \quad \hat{\alpha}_p < \alpha_p + 2\alpha_\infty\, \alpha_p \left\{ \left| \hat{\Phi}(0) \right| + \left| \hat{\Phi}(1) \right| \right\}$$

Proof. The inequality (3.9) is a direct consequence of (3.5) while (3.10) follows immediately from the relation

$$\hat{\Phi}(t) = G(t,0)\, \hat{\Phi}(0) - G(t,1)\, \hat{\Phi}(1)$$

which can easily be verified from the definition of G. The remaining inequalities can now be obtained in a simple manner from (3.8), (3.4), (3.9) and (3.10). ‡

It is of interest to view Lemma 3.1 in the following manner. Suppose that the boundary conditions (3.2) are such that the problems (3.1),(3.2) is as well conditioned as possible (i.e. α_∞ is as small as possible).

If the problem at hand is (3.1),(3.3) and $|\hat{\Phi}(0)| + |\hat{\Phi}(1)|$ is of moderate size, then we can be assured that the condition number $\hat{\alpha}_\infty$ is at worst $O(\alpha_\infty^2)$. Thus, if α_∞ is of moderate size then so is $\hat{\alpha}_\infty$.

We now show that it is possible to construct separated boundary conditions (i.e. rank $(\hat{B}_0)$ + rank $(\hat{B}_1) = n$) such that $\hat{\alpha}_\infty$ is at worst $O(\alpha_\infty^2)$.

Lemma 3.2 (de Hoog and Mattheij [3]). **There exist separated boundary conditions $\hat{B}$ of the form (2.3) such that**

$$\hat{\alpha}_p < \alpha_p + 4\,\alpha_\infty\,\alpha_p \,.$$

Proof Consider the singular value decomposition

$$Y(1) = \Phi(1)\,\Phi^{-1}(0) =: \quad U\,D\,V^T$$

where U, V are orthogonal matrices and

$$D = \mathrm{diag}\,(1/d_1, \ldots, 1/d_r, d_{r+1}, \ldots, d_n)$$

with $0 < d_j < 1$. Now define

$$P_0 = \left[\begin{array}{c|c} 0 & 0 \\ \hline 0 & I_{n-r} \end{array}\right], \qquad P_1 = I - P_0$$

$$D_0 = \text{diag } (d_1, \ldots, d_r, 1, \ldots, 1)$$

$$D_1 = \text{diag } (1, \ldots, 1, d_{r+1}, \ldots, d_n)$$

$$\hat{B}_0 = P_0 V^T \quad \text{and} \quad \hat{B}_1 = P_1 \, U^T.$$

It is easy to verify that

$$\hat{\Phi} = Y \; V \; D_0$$

and hence

$$|\hat{\Phi}(0)| = |VD_0| < 1$$

$$|\hat{\Phi}(1)| = |UDV^T \; V \; D_0| = |U \; D_1| < 1 \; .$$

The result now follows from (3.12). #

4. An Analysis of Shooting Methods. When the size of the coefficient matrix A is moderate, the analysis of finite difference schemes applied to (1.1), (1.2) is straightforward. In fact for many schemes the condition numbers of the finite difference schemes will converge to the condition numbers β and α_p (see for example de Boor, de Hoog and de Keller [2]). Then the usual argument of consistency and stability yields convergence. However for some problems it is not practical to assume that $h\|A\|_\infty$ (h is the maximum step size) is small and the analysis required to show convergence is much more delicate (see for example the analyses in Weiss [6] and Ascher and Weiss [1] for symmetric collocation schemes applied to singular perturbation problems).

In this section we analyse shooting techniques with the view of obtaining sufficient conditions for convergence of numerical schemes for problems where $\|A\|_\infty >> 1$.

We begin with simple shooting for the problem

(4.1) $\quad L\,y = f\,, \quad 0 < t < 1\,; \qquad B\,y = b$

and consider two associated boundary value problems

(4.2a) $\quad L\,\hat{\Phi} = 0\,, \quad \hat{B}\,\hat{\Phi} = I$

(4.2b) $\quad L\,\hat{y} = f\,, \quad \hat{B}\,\hat{y} = 0$

Our motivation is that numerical schemes applied to (4.2 a,b) may be easier to analyse than if the scheme were applied directly to (4.1). We have in mind here the situation where the boundary conditions $\hat{B}$ are substantially simpler than B (for example we could have separated boundary conditions for $\hat{B}$).

As previously we write

$$y = \hat{\Phi}c + \hat{y}$$

and on imposing the boundary condition $B\,y = b$ we obtain

$$(B\,\hat{\Phi})c = b - B\,\hat{y} \,.$$

Thus,

$$(4.3) \quad y = \hat{\Phi}\,(B\,\hat{\Phi})^{-1}\,(b - B\,\hat{y}) + \hat{y} \,.$$

Now suppose that we have approximations

$$\hat{\Psi} \sim \hat{\Phi} \,, \quad \hat{z} \sim \hat{y} \,.$$

Define

$$\hat{E} := \hat{\Psi} - \hat{\Phi} \,, \quad \hat{e} := \hat{z} - \hat{y}$$

and

$$(4.4) \quad \hat{y} := \hat{\Psi}\,(B\,\hat{\Psi})^{-1}\,(b - B\,\hat{z}) + \hat{z}$$

where we have assumed that $B\,\hat{\Psi}$ is nonsingular.

The relevance of $\hat{y}$ is that it will be the approximation to y of a scheme applied to (4.1) if the same scheme applied to (4.2 a,b) yields approximations $\hat{\Psi}$ and $\hat{z}$ to $\hat{\Phi}$ and $\hat{y}$ respectively. The following lemma gives a bound on the error $|\hat{y} - y|$.

Lemma 4.1 Let $\hat{y}$, $\hat{E}$ and $\hat{e}$ be defined as above. If

$$|\hat{B}\,\Phi|\;|B\,\hat{E}| < 1,$$

then $B\,\hat{\Psi}$ is nonsingular and

$$|y - \tilde{y}| < \frac{1}{1-|\hat{B}\,\Phi|\,|\hat{B}\,E|}\{|\Phi|\,[|B\,\hat{E}|\,|\hat{B}\,y| + |B\,\hat{e}|]$$

$$+\,|\hat{E}|\;[|\hat{B}\,y| + |\hat{B}\,\Phi|\,|B\,\hat{e}|]\,\} + |\hat{e}|$$

<u>Proof</u> Since from (3.4),

$$(B \hat{\Phi})^{-1} = \hat{B} \Phi$$

we have

$$B \hat{\Psi} = B \hat{\Phi} + B \hat{E}$$

$$= B \hat{\Phi} (I + \hat{B} \Phi B \hat{E})$$

and as $B \hat{\Phi}$ is nonsingular, $B \hat{\Psi}$ is also nonsingular if $| \hat{B} \Phi || B \tilde{E}| < 1$.

With

$$\Psi := \hat{\Psi}(B \Psi)^{-1}$$

it follows from (4.3) and (4.4) that

$$(4.5) \quad |\hat{y} - y| \leq |(\Phi - \Psi) (b - B \hat{y})|$$

$$+ |\Psi|| B \hat{e}| + |\hat{e}|$$

We now bound each of the terms on the right hand side of this inequality. From (4.3),

$$\hat{B}\, y = (B\, \hat{\Phi})^{-1}(b - B\, \hat{y})$$

and hence

$$\left|(\Phi - \Psi)\ (b - B\, \hat{y})\right|$$

$$\leq \left|(\Phi - \Psi)\ B\, \hat{\Phi}\right| \left|\hat{B}\, y\right|$$

$$= \left|\hat{\Phi} - \hat{\Psi}(B\, \hat{\Psi})^{-1} B\, \hat{\Phi}\right| \left|\hat{B}\, y\right|$$

$$\leq \left|\hat{\Phi}\ (I - (B\, \hat{\Phi} + B\, \hat{E})^{-1}\ B\, \hat{\Phi})\right| \left|\hat{B}\, y\right|$$

$$+ \left|\hat{E}\right| \left|(B\, \hat{\Phi} + B\, \hat{E})^{-1}\ B\, \hat{\Phi}\right| \left|\hat{B}\, y\right|$$

$$= \left|\Phi\, B\, \hat{E}\ (I + (B\, \hat{\Phi})^{-1}\ B\, \hat{E})^{-1}\right| \left|\hat{B}\, y\right|$$

$$+ \left|\hat{E}\right| \left|(I + (B\, \hat{\Phi})^{-1}\ B\, \hat{E})^{-1}\right| \left|\hat{B}\, y\right|$$

$$\leq \frac{\left|\Phi\right| \left|B\, \hat{E}\right| \left|\hat{B}\, y\right| + \left|\hat{E}\right| \left|\hat{B}\, y\right|}{1 - \left|\hat{B}\, \Phi\right| \left|B\, \hat{E}\right|}$$

Also,

$$|\Psi| \, |B \hat{e}|$$

$$= \left|(\hat{\Phi} + \hat{E}) \, (B \hat{\Phi})^{-1} \, (I + B \hat{E} \, (B \hat{\Phi})^{-1})^{-1}\right| \, |B \hat{e}|$$

$$< \frac{(|\Phi| + |\hat{E}| \; |\hat{B} \Phi|) \, |B \hat{e}|}{1 - |\hat{B} \Phi| \; |B \hat{E}|}.$$

and the result now follows on substitution of these inequalities in (4.5). #

<u>Corollary 4.1</u> Let y, $\hat{E}$ and $\hat{e}$ be defined as in Lemma 4.1,

$$\|\hat{E}\|_\infty = \varepsilon_1 \, , \quad \|\hat{e}\|_\infty = \varepsilon_2$$

and

$$\max\{|B_0|, \; |B_1|\} = \max\{|\hat{B}_0|, \; |\hat{B}_1|\} = 1.$$

Then,

$$\|\hat{y} - y\|_\infty < \frac{2}{1-4\beta\varepsilon_1} \{(2\beta+1)\|y\|_\infty \varepsilon_1 \cdot$$

$$+ \beta(1 + 2\varepsilon_1) \varepsilon_2\} + \varepsilon_2 .$$

Proof. The result is an immediate consequence of Lemma 4.1.

#

It is interesting to note that the estimate in Corollary 4.1 only involves the conditioning of problem (4.1). Basically the estimates confirm the fact that if a numerical scheme yields accurate estimates (in an absolute rather than relative sense) to the solutions of (4.2 a,b), then the scheme will also yield a good approximation to the solution of (4.1) provided that the condition numbers of this problem are moderate. However we shall defer further discussion of the above result until we have analysed the more general case of multiple shooting.

The basic idea of multiple shooting is to break the interval into a number of subintervals on which different representations of the solution are used. Specifically, let

$$0 = t_0 < t_1 < \dots < t_N = 1$$

and define

$$\hat{B}_j g := \lim_{t \to t_{j-1}^+} \hat{B}_{0j} g(t) + \lim_{t \to t_j^-} \hat{B}_{1j} g(t) .$$

On the interval $t_{j-1} < t < t_j$, $j = 1,..,N$ we represent the solution of (4.1) as

$$y = \hat{\Phi}_j c_j + \hat{y}_j , \quad t_{j-1} < t < t_j, \; j = 1,..,N$$

where

$$\text{(4.6a)} \quad L\,\hat{\Phi}_j = 0 \;,\; t_{j-1} < t < t_j,\; \hat{B}_j \hat{\Phi}_j = I$$

$$\text{(4.6b)} \quad L\,\hat{y}_j = f \;,\; t_{j-1} < t < t_j,\; \hat{B}_j \hat{y}_j = 0 .$$

In order to determine the vectors c_j, $j = 1,..,N$ we now impose continuity at t_j, $j = 1,..,N-1$ and the boundary condition $B\,y = b$. This yields the system of equations

$$\hat{\Phi}_{j+1}(t_j)\, c_{j+1} - \hat{\Phi}_j(t_j)\, c_j = d_j \;, \quad j = 1,...,N-1$$

$$B_0 \hat{\Phi}_1(0)\, c_1 + B_1 \hat{\Phi}_N(1)\, c_N = d_N$$

where

$$d_j = \hat{y}_j(t_j) - \hat{y}_{j+1}(t_j) \ , \ j = 1,\ldots,N-1$$

$$d_N = b - B_0\hat{y}_1(0) - B_1\hat{y}_N(1)$$

which we write in matrix form as

$$M\underset{\sim}{c} = \underset{\sim}{d} \ .$$

It turns out that an explicit solution to this system can be found. Specifically,

$$(4.7) \qquad c_j = \hat{B}_{1j}\left(\sum_{k=1}^{N-1} G(\cdot, t_k)d_k + \Phi(\cdot)d_N\right)$$

and this expression can be deduced from the inverse of the usual multiple shooting matrix (i.e. when $\hat{B}_{0j} = I$, $\hat{B}_{1j} = 0$) in [5]. Alternatively, (4.7) may be deduced from the observation that the problem

$$L\,u = 0 \ , \ t \in \bigcup_{k=1}^{N} (t_{k-1}, t_k)$$

with the jump conditions

$$\lim_{t \to t_j^+} u(t) - \lim_{t \to t_j^-} u(t) = d_j \quad , \; j = 1,\ldots,N-1$$

and boundary conditions

$$B\, u = d_N$$

has the solution

$$u = \sum_{k=1}^{N-1} G(\cdot, t_k) d_k + \Phi\, d_N \; .$$

Suppose now that we have approximations $\hat{\Psi}_j \sim \hat{\Phi}_j$, $\hat{z}_j \sim \hat{y}_j$, $j = 1,\ldots,N$ which we use to construct an approximate solution

$$\hat{y} = \hat{\Psi}_j\, \hat{c}_j + \hat{z}_j \quad , \; t_{j-1} < t < t_j \quad , \; j = 1,\ldots,N$$

to (4.1) by imposing continuity and the boundary condition $B\, y = b$. This yields the system of equations

$$\hat{\Psi}_{j+1}(t_j)\, \hat{c}_{j+1} - \hat{\Psi}_j(t_j)\, \hat{c}_j = \hat{d}_j \quad , \; j = 1,\ldots,N-1$$

$$B_0\, \hat{\Psi}_0(0) + B_1 \hat{\Psi}_N(1)\, \hat{c}_N = \hat{d}_N$$

where

$$\hat{d}_j = \hat{z}_j(t_j) - \hat{z}_{j+1}(t_j) \; , \; j = 1,\ldots,N-1$$

$$\hat{d}_N = b - B_0 \, \hat{z}_1(0) - B_1 \, \hat{z}_N(0)$$

which we write as

$$\hat{M} \, \hat{\underset{\sim}{c}} = \hat{\underset{\sim}{d}} \; .$$

If $\hat{M}$ is nonsingular, then the approximation $\hat{y}$ to y is well defined. An analysis similar to that given in Lemma 4.1 now yields.

<u>Lemma 4.2</u> Let

$$\max\{|B_0|, \; |B_1|\} = \max\{|B_{0j}|, |B_{1j}|\} = 1, \; j = 1,\ldots,N$$

$$|\hat{\Psi}_j(t)-\hat{\Phi}_j(t)|<\varepsilon_1, |\hat{z}_j(t)-\hat{y}_j(t)|<\varepsilon_2, \; t_{j-1}\leq t\leq t_j ; j=1,\ldots,N$$

and

$$4(N+1) \; \alpha_\infty \; \varepsilon_1 < 1 \; .$$

Then, $\hat{M}$ is nonsingular and

$$(4.8) \quad \|\hat{y} - y\|_\infty < \frac{2}{1-4(N+1)\alpha_\infty \varepsilon_1} \{(1 + 2(N+1)\alpha_\infty) \|y\|_\infty \varepsilon_1$$

$$+ (N+1) \alpha_\infty (1 + 2 \varepsilon_1) \varepsilon_2\} + \varepsilon_2 .$$

Again we see that the only condition number involved in the estimate is one associated with the problem (4.1). Thus, if a numerical scheme applied to the problems (4.6 a,b) yields accurate approximations, then the scheme applied to (4.1) will also yield a good approximation provided that α_∞ is of moderate size. It should be recalled from Lemma 2.2 that on each subinterval (t_{j-1}, t_j), boundary conditions $\hat{B}_j$ exist such that the problem on the subinterval is at least as well conditioned as problem (4.1). Furthermore, if α_∞ is of moderate size, then from Lemma 3.2 there exist separated boundary conditions $\hat{B}_j$ such that the problems (4.6 a,b) are still well conditioned.

The estimate (4.8) is quite useful in the analysis of numerical schemes for singular perturbation problems with turning points. If we take the turning points to be $t_1,\ldots,t_{N-1}$, then the problems (4.6 a,b) are singular perturbation problems with boundary layers. Thus if we can establish convergence of the numerical schemes for these boundary layer problems, then (4.8) guarantees convergence of the scheme applied to (4.1). In other words we can restrict attention to the analysis of schemes applied to boundary layer problems and this substantially simplifies the notation required.

To demonstrate this, consider the problem

$$(4.9) \qquad \varepsilon u'' + t u' = g , \quad -1 < t < 1$$

$$u(-1) = c , \quad u(1) = a$$

Then, with

$$y_1 = u , \quad y_2 = \sqrt{\varepsilon}\ u'$$

we obtain the first order system

$$(4.10a) \qquad L y = y' - Ay = f , \quad -1 < t < 1$$

$$(4.10b) \qquad B y = B_0 y(-1) + B_1 y(1) = b$$

where

$$A = \begin{bmatrix} 0 & 1/\sqrt{\varepsilon} \\ 0 & -t/\varepsilon \end{bmatrix} , \quad f = \begin{bmatrix} g/\sqrt{\varepsilon} \\ 0 \end{bmatrix}$$

$$B_0 = \begin{bmatrix} 0 & 0 \\ 1 & 0 \end{bmatrix} , \quad B_1 = \begin{bmatrix} 1 & 0 \\ 0 & 0 \end{bmatrix} , \quad b = \begin{bmatrix} a \\ c \end{bmatrix}$$

Note that in contrast to (4.1), we now have a problem defined on $[-1,1]$ rather than $[0,1]$. Although we can easily transform the problem to $[0,1]$, this has not been done since (4.9) is a popular example which is usually given on the larger interval. It is easy to verify that the condition number α_∞ associated with (4.10 a,b) is independent of ε and is $O(1)$. Furthermore as (4.10 a,b) has a turning point at $t = 0$ we divide the interval $(-1,1)$ into two parts by taking $t_0 = -1$, $t_1=0$ and $t_2=1$. On these intervals we now take

$$\hat{B}_1 y = \lim_{t \to 0^-} y(t)$$

$$\hat{B}_2 y = \lim_{t \to 0^+} y(t) .$$

Thus (4.6 a,b) are final and initial value problems for $j = 1$ and $j = 2$ respectively and it can be verified that the condition number associated with these subproblems are $O(1)$. Suppose we now wish to analyse the performance of the centred Euler (Box) scheme applied to (4.10 a,b) on a grid that has been chosen such that the internal layer about $t = 0$ is adequately resolved. Provided that $t = 0$ is a grid point (this assumption is only made for simplification), convergence can be established by examining the convergence of the centred Euler scheme to the final and initial value problems (4.6 a,b). The estimate (4.8) then immediately yields convergence of the scheme applied to (4.10 a,b).

As a further illustration of the utility of the estimate (4.8) consider the centred Euler scheme applied to (4.1) with

$$A = \frac{1}{\varepsilon} \begin{bmatrix} \cos 2\beta t & -\sin 2\beta t \\ -\sin 2\beta t & -\cos 2\beta t \end{bmatrix}$$

and

$$B_0 = \begin{bmatrix} 0 & 0 \\ 0 & 1 \end{bmatrix}, \qquad B_1 = \begin{bmatrix} \cos\beta & -\sin\beta \\ 0 & 0 \end{bmatrix}$$

When ε is small, this problem has boundary layers at $t = 0$ and $t = 1$. If these layers are adequately resolved (see for example Weiss [6]), the analysis of Weiss [6] establishes convergence of the scheme provided $\beta \neq (2k+1)\pi/2$ where k is an integer. Let us now consider the case when $\beta = (2k+1)\pi/2$ and ε is small. As in the previous example divide the interval into two parts by taking $t_0 = 0$, $t_1 = 1/2$ and $t_2 = 1$. Now define

$$\hat{B}_{01} = B_0 , \quad \hat{B}_{12} = B_1$$

$$\hat{B}_{11} = \begin{bmatrix} \cos\beta/2 & -\sin\beta/2 \\ 0 & 0 \end{bmatrix}, \quad \hat{B}_{02} = \begin{bmatrix} 0 & 0 \\ \sin\beta/2 & \cos\beta/2 \end{bmatrix}$$

and introduce a grid with an additional region of resolution about $t = 1/2$ (with one of the grid points at $t = 1/2$). The analysis of Weiss [6] now shows that the centred Euler scheme converges for the subproblems (4.6 a,b) and hence (4.8) ensures convergence of the scheme applied to (4.1). Thus, in this case, the scheme can be salvaged by the introduction of a grid with an artificial layer of resolution in the interior of the interval. In fact this technique can be generalised to include the more general problems addressed in [6].

5. Concluding Remarks

The results of the analysis presented in this paper are in some sense intuitively obvious. However, by presenting them in a rigorous fashion we are able to deduce sufficient conditions for convergence of numerical schemes that are relatively simple to verify. Although we have concentrated on convergence, stability of the schemes can be approached in a similar manner.

References

1. Ascher, U. and Weiss, R. Collocation for singular perturbation problems II : Linear first order systems without turning points. Math. Comp 43 (1984), 157-187.
2. de Boor, C., de Hoog, F. and de Keller, H. The stability of one step schemes for first-order two-point boundary value problems, SIAM J. Numer. Anal., 20 (1983), 1139-1146.
3. de Hoog, F. and Mattheij, R., On dichotomy and well conditioning in boundary value problems, submitted for publication.

4. Keller, H.B., Numerical solution of two-point boundary value problems, SIAM regional conference series, 24, Philadelphia, 1976.

5. Lentini, L., Osborne, M.R. and Russell, R.D., The close relationships between methods for solving two-point boundary value problems, to appear in SINUM.

6. Weiss, R., An analysis of the box and trapezoidal schemes for linear singularly perturbed boundary value problems, Math. Comp 42 (1984), 41-67.

Progress in Scientific Computing, Vol. 5
Numerical Boundary Value ODEs

ON NON-INVERTIBLE BOUNDARY VALUE PROBLEMS

R.M.M. Mattheij
Mathematisch Instituut
Katholieke Universiteit
6525 ED Nijmegen
The Netherlands

F.R. de Hoog
Division of Mathematics & Statistics
CSIRO
G PO Box 1965
Canberra
Australia

Abstract

For non-invertible boundary value problems, i.e. where the boundary conditions as such do not determine the solution uniquely, the usual concepts of condition numbers and stability do not apply. Such problems typically arise when the interval is semi-infinite. If one assumes that the desired solution is bounded the boundary conditions are sufficient to give a unique solution (as an element of the bounded solutions manifold). Another type of problems are eigenvalue problems, where both the dynamics and the boundary conditions are homogeneous. We shall introduce sub-condition numbers that indicate the sensitivity of the problem with respect to perturbations of a relevant sub-problem. We also discuss a numerical method that computes such sub-condition number to demonstrate its applicability. Finally we give a number of numerical examples to illustrate both the theory and the computational method.

§1. Introduction

Consider the differential equation

$$(1.1)\qquad \frac{dx}{dt} = L(t)x+f(t), \quad 0 \le t < \infty,$$

where $L(t)$ is an nth order matrix for all t and $x(t)$ and $f(t)$ are n dimensional vectors. Let x satisfy the boundary condition (BC)

$$(1.2)\qquad M_0\, x(0) + M_1\, x(\infty) = b,$$

where M_0 and M_1 are nth order matrices and $M_1\, x(\infty) := \lim_{t\to\infty} M_1\, x(t)$ is assumed to exist. We seek a solution of (1.1) and (1.2) that is bounded on $[0,\infty)$. A trivial case arises when $M_1 = 0$, M_0 is nonsingular, f is bounded and the zero solution of $\frac{dx}{dt} = Lx$ is asymptotically stable; this is a stable initial value problem. If, more generally, the homogeneous system also possesses unstable modes, the boundedness requirement for x poses a so called <u>conditionally stable problem</u>; that is we seek a solution on the <u>stable manifold</u>. Apparently the BC (1.2) should provide sufficient conditions to uniquely define a solution on this manifold. On the other hand, viewed as a BVP on $[0,\infty)$, (1.1) and (1.2) alone will not uniquely define a solution. Therefore we may say that the problem as such is <u>non-invertible</u>, cf [11].

Another class of non-invertible problems is the determination of a solution x for a special value of a parameter λ, the <u>eigenvalue</u>, such that

$$(1.3)\qquad \frac{dx(t,\lambda)}{dt} = L(t,\lambda)\, x(t,\lambda)$$

$$(1.4)\qquad M_0\, x(0,\lambda) + M_1\, x(1,\lambda) = 0.$$

It is immediately clear that, if there exists a solution $x(.,\lambda)$, then also any multiple will be a solution ("eigensolution"). Hence, again there is no uniqueness. Of course

we may also have a combination of both types of non-invertible problems (eigenproblem on $[0,\infty)$.
Problems of the kind (1.1), (1.2) and (1.3), (1.4) have extensively been studied in literature, cf [1,8,9,10,11,12, 13,22]. Typical for most approaches is that approximating invertible BVP are constructed one way or another, except for [11] where a least squares solution is developed and [22], which in a way comes close to our stable manifold approximation idea, to be outlined in sections 2, 4 and 5.
In constructing an invertible BVP on $[0,\infty)$ one has to assume that the system matrix $L(t)$ approaches a limit-value $(t \to \infty)$. This condition, though meaningful in many applications, seems unnecessarily restrictive. In our presentation we rather like to view the problem from the angle of solutions spaces. In this the notion of dichotomy provides for a separation of the "bounded manifold" and the "unstable manifold", cf [3,6]. Roughly speaking a fundamental solution Y is dichotomic, if it can be split into parts of increasing and decreasing (or nonincreasing) solutions. On a finite interval $[0,1]$ say, it is then meaningful to normalize the increasing solutions at $t = 1$ and the decreasing solutions at $t = 0$, such that $\max_{t\in(0,1)} ||Y(t)|| = 1$. By this normalization the condition number measuring the sensitivity of the solution with respect to the BC, viz.

$$(1.5) \qquad \beta := \max_t ||Y(t)Q^{-1}||,$$

where

$$(1.6) \qquad Q := M_0\, Y(0) + M_1\, Y(1),$$

can then be estimated by (cf [18])

$$(1.7) \qquad \gamma := ||Q^{-1}||.$$

As turned out in e.g. [6,17] the latter quantity is also decisive in estimating the condition number, measuring the sensitivity of the solution with respect to perturbations in the forcing term of (1.1), viz.

$$(1.8) \qquad \alpha := \max ||G(t,s)||,$$

where the Green's function is defined by

$$(1.9) \qquad G(t,x) := \begin{cases} Y(t)Q^{-1}M_0Y(0)Y^{-1}(s), & t > s \\ -Y(t)Q^{-1}M_1Y(1)Y^{-1}(s), & t < s. \end{cases}$$

However, in the non-invertible case, where Q^{-1} does not exist, these concepts have to be adapted. We introduce similar stability constants, sub-condition numbers, for the conditionally stable problem in section 2 and for eigenvalue problems in section 3. Finally we discuss a numerical algorithm based on multiple shooting which computes these subcondition numbers in section 5. Their usefulness is demonstrated in section 6.

§2. Conditioning of problems on infinite intervals

In this section we investigate the problem (1.1), (1.2). As has been shown elsewhere in this proceedings in dealing with BVP an important notion is dichotomy (cf [3 ,14]). Here we shall use the following particular

Definition 2.1. The fundamental solution Y of (1.1) is called dichotomic if there exists constants κ and $\lambda < 0$, and a projection P_1, such that

(i) $||Y(t)P_1Y^{-1}(s)|| \leq \kappa \exp(\lambda(s-t))$, $t < s$,

(ii) $||Y(t)(I-P_1)Y^{-1}(s)|| \leq \kappa$, $t > s$.

($||.||$ denotes any associated norm).

If we assume dichotomy then the following existence result for a bounded solution y can be given (cf. [3]).

Theorem 2.2. Let L denote the Banach space of all vector functions f which are integrable on $\mathbb{R}_+$ with norm $|f| := \int_0^\infty |f(t)|dt$. Then (1.1) has at least one bounded solution y_b for every $f \in L$ iff Y is dichotomic.

If there is dichotomy a bounded solution is given by

$$(2.3) \qquad y_b(t) = \int_0^t Y(t)(I-P_1)Y^{-1}(s)f(s)ds - \int_t^\infty Y(t)P_1Y^{-1}(s)f(s)ds.$$

We intend to report on existence of solutions and well-conditioning related to existence of dichotomy elsewhere. For our purpose it now suffices to assume that some bounded solution, y_b exists. The desired particular solution y then lies in the linear manifold $y_b \oplus \text{span}(Y(I-P_1))$.

The most widely used method to compute the solution y in practice employs the latter fact in that it tries to determine a point β large enough such that a suitable "replacing" terminal condition can be used. In this way the problem is reduced to a BVP on a finite interval and can be solved by any standard BVP algorithm, cf [1,8,9,12,13,22]. This idea seems to work well if the systems L(t) approaches a limit, whence the "direction" of the unstable subspace approaches a limit. For more general differential equations, notably where there is a "directional activity" (i.e. the solutions rotate) everywhere (cf [24]), it may not work, which we show first.

Let us introduce a normalized fundamental solution F_T say, for any $T > 0$ (cf. section 1)

$$(2.4)\qquad F_T(t) = Y(t)P_1Y^{-1}(T)+Y(t)(I-P_1)Y^{-1}(0).$$

Now consider a finite BC

$$(1.2)'\qquad M_0\, x(0) + M_T\, x(T) = b\ ,$$

which is assumed to be such that (1.1), (1.2)' is a well-posed problem, as is the case in standard approaches, cf. [1,8,9,12,13]. A natural stability constant for this BVP on [0,T] would be (cf. (1.5))

$$(2.5)\qquad \beta_T := \max_t \|F_T(t)[M_0F_T(0)+M_TF_T(T)]^{-1}\|,$$

and for its estimate (cf (1.7))

$$(2.6)\qquad \gamma_T := \|[M_0F_T(t)+M_TF_T(T)]^{-1}\|.$$

If we would somehow choose a fixed matrix M_T (for all T) then we may not have a limiting behaviour of the stability constants, see

<u>Example 2.7</u>. Let $F_T(t) := \begin{bmatrix} \cos t & \sin t \\ -\sin t & \cos t \end{bmatrix} \begin{bmatrix} e^{10(t-T)} & 0 \\ 0 & e^{-10t} \end{bmatrix}$

and let $M_0 = M_T = \begin{bmatrix} 1 & 0 \\ 0 & 1 \end{bmatrix}$. Then

$$M_0F_T(0)+M_TF_T(T) = \begin{bmatrix} e^{-10T}+\cos T & e^{-10T}\sin T \\ -\sin T & 1+e^{-10T}\cos T \end{bmatrix}.$$

If, for some integer k, $T = (k+\frac{1}{2})\pi$, $\gamma_T \approx e^{10T}$, but if e.g. $T = k\pi$ then $\gamma_T \approx 1$. □

In the example above only the choice $M_T = 0$ would avoid this "oscillatory growth" of γ_T as $T \to \infty$; however, $M_T = 0$ implies $\gamma_T \sim e^{10T}$ (we need some terminal condition for the nonreduced problem, of course, due to the presence of the unstable mode $\binom{\cos t}{-\sin t}e^{10t}$). We conclude that only a choice of M_T that is dependent on T but of which the last row does not make a small angle with the vector $\binom{\cos T}{-\sin T}$ (cf [18]) may work. We are not aware of any method that is designed to acomplish this.

Therefore we investigate instead the possibility to determine the stable manifold somehow and to use this in combination with the BC. For such an approach the stability and conditioning concepts have to be adapted. This will be done next.

Define the "nongrowing part" of the fundamental solution as

(2.8) $\quad F^2(t) := Y(t)(I-P_1)Y^{-1}(0),$

and consequently a condition-matrix (cf. (1.6))

(2.9) $\quad Q_T := M_0F^2(0) + M_1F^2(T).$

Suppose we have a bounded solution y_b (cf. 2.2), then we look for a vector c_T "satisfying"

(2.10) $\quad Q_T^2c_T = M_0y_b(0) + M_1y_b(T),$

with

(2.11 a) $\quad \lim_{T\to\infty} c_T = c$

where c is such that

(2.11 b) $\quad x(t) = y_b + F^2(t)c.$

(NB the vector c in (2.11 b) exists by assumption). We have put quotation marks around satisfying as (2.10) may not be exactly true for finite T. One should realize however, that the vector c, defined by (2.11 b) is not necessarily unique, but $(I-P_1)Y^{-1}(0)c$, i.e. its projection on the stable solution space is unique. This naturally induces a projection principle to solve (2.9), viz. by least squares, anticipating a zero residual in the limit, $(T \to \infty)$. Denoting the Moore-Penrose generalized inverse (cf. [5, p.139]) of Q_T^2 by $[Q_T^2]^+$, we are thus led to consider the following <u>subcondition</u> number, cf (2.5)

(2.12) $$\bar{\beta}_T := \max_t ||F^2(t)[Q_T^2]^+||$$

and its estimate (cf (2.6))

(2.13) $$\bar{\gamma}_T := ||[Q_T^2]^+||.$$

One should realize that $[Q_T^2]$ is expected to be a rank (n-k) matrix for all T and is anticipated to have the same kernel for all T too. This makes it meaningful to investigate the existence of $\lim_{T\to\infty} \bar{\beta}_T$ and $\lim_{T\to\infty} \bar{\gamma}_T$:

<u>Theorem 2.14</u>. If $\lim_{T\to\infty} [M_0F^2(0)+M_1F^2(T)]$ exists and has rank (n-k), then $\lim_{T\to\infty} [Q_T^2]^+$ exists.

<u>Proof</u>. It is simple to see that for T large enough rank (Q_T^2) = n-k. Since $Q^2 := \lim_{T\to\infty} Q_T^2$ exists it follows from Cor. 3.5 in [25] that $\lim [Q_T^2]^+$ exists. □

<u>Property 2.15</u>. $\lim_{T\to\infty} [M_0F^2(0)+M_1F^2(T)]$ exists if

(i) $M_1 = 0$

or

(ii) Y is exponentially dichotomic (that is instead of (2.1) (ii): $||Y(t)(I-P_1)Y^{-1}(s)|| \le \kappa \exp(\mu(t-s))$, $\mu < 0$, $t > s$)

or

(iii) assuming F^2 can be split into $F^2(I-S)+F^2S$, where S is a projection such that $||F^2(t)(I-S)F^2(s)|| \le \kappa \exp(\mu(t-s))$, $\mu < 0$, $t > s$ and $||F^2(t)SF^2(S)|| \ge \tilde{\kappa}$, for some $\tilde{\kappa} > 0$, this splitting is such that $\lim_{T\to\infty} F^2(T)S$ exists.

The proof is straightforward. Note that (iii) implies that the nondecaying-nonincreasing part of the fundamental

solution has asymptotically constant directions. All other solutions, whether increasing or decreasing may have varying directions as $T \to \infty$.
It is well-known that a suitable generalized inverse can be given in terms of the SVD (Singular Value Decomposition, cf. [5]): There exist orthogonal matrices U,V and $\Sigma = \mathrm{diag}(\sigma_1,...,\sigma_n)$ a semipostive diagonal matrix, such that

$$Q^2 = U\,\Sigma\,V^T. \tag{2.16}$$

Then

$$||[Q^2]^+||_2 = \min_{\sigma_j \neq 0} \sigma_j^{-1}. \tag{2.17}$$

Now assume for simplicity $Y(0) = I_n$ and

$$P_1 = \begin{bmatrix} I_k & \emptyset \\ \emptyset & \emptyset \end{bmatrix}. \tag{2.18}$$

Then we can also give (adapted) "Green's functions" (cf (1.9)) as follows. Let U,V,Σ be such that $\sigma_1,...,\sigma_k = 0$. We then obtain

$$[Q^2]^+[M_0F^2(0)+M_1F^2(\infty)] = I-P_1. \tag{2.19}$$

This induces an obvious analogue of (1.9):

$$[G(t,s)]^2 = \begin{cases} Y(t)[Q^2]^+M_0Y(0)(I-P_1)Y^{-1}(s), & t > s \\ Y(t)[Q^2]^+M_1Y(\infty)(I-P_1)Y^{-1}(s), & t < s \end{cases} \tag{2.20}$$

and leaving out the first k zero columns. Such Green's functions naturally give rise to stability constants for computations dealing with the stable manifold only.

We conclude this section with a remark on so called "boundary value methods" for initial value problems. In such methods an IVP is formulated as a BVP, using certain terminal conditions. Quite often these terminal conditions are satisfied only asymptotically, which makes such a method resemble infinite interval problems. However, given the more complicated nature of BVPs as compared to IVPs - in particular the conditioning of the problem (and the related dichotomy) - such an approach, cf [26], might be less attractive for less trivial problems i.e. when there is a significant directional activity. In such situations techniques aimed more directly at computing solutions of the stable manifold as such (cf [18, 22]) should then be preferred.

§3. Conditioning of eigenvalue problems

Let $Y(t,\lambda) =: Y(t)$ be a fundamental solution of (1.3), (1.4), which is dichotomic as in 2.1 with $\lambda = 0$. Then define

$$F(t) = Y(t)(I-P_1)Y^{-1}(0) + Y(t)P_1Y^{-1}(1). \tag{3.1}$$

This normalized fundamental solution gives rise to a condition-matrix

$$Q = M_0F(0) + M_1F(1). \tag{3.2}$$

Since λ is assumed to be an exact eigenvalue there exists some vector c such that

$$x(t) = F(t)c, \tag{3.3}$$

where the nontrivial c satisfies

$$Qc = 0. \tag{3.4}$$

Of course rank (Q) should be $\le n-1$.
The best way to characterize the kernel of Q is employing singular vectors. Hence we we again use the SVD,

$$Q = U\Sigma V^T, \tag{3.5}$$

where in $\Sigma = \text{diag}(\sigma_1,\ldots,\sigma_n)$ we assume $\sigma_{i+1} \ge \sigma_i$, $i \ge 1$. In particular we have $\sigma_n = 0$. For simplicity we let $\sigma_{n-1} > 0$ (so rank $(Q) = n-1$). Let the last column of V be denoted by v_n, then "the" solution of (3.4) is given by

$$c = v_n. \tag{3.6}$$

(Of course any scalar multiple of v_n is a solution as well.)

We now like to give a sensitivity analysis of this problem. If we perturb M_0 and M_1 by δM_0 and δM_1 respectively we obtain a (possibly nonsingular) condition-matrix

$$(3.7) \qquad Q+\delta Q := (M_0+\delta M_0)F(0)+(M_1+\delta M_1)F(1)$$

We proceed as follows: First we remark that it makes sense, from a practical point of view, to solve the perturbed problem by seeking a vector $c+\delta c$ such that

$$(3.8) \qquad (Q+\delta Q)(c+\delta c) = 0$$

in a "least squares sense". This precisely means that we ask for the direction of a singular vector of $(Q+\delta Q)$ corresponding to the smallest singular value. So let

$$(3.9) \qquad (Q+\delta Q) = (U+\delta U)(\Sigma+\delta\Sigma)(V+\delta V)^T,$$

with $\delta\Sigma =: \mathrm{diag}(\delta\sigma_1,\ldots,\delta\sigma_n)$; let the last column of δV be denoted by δv_n. Then we may identify δc with δv_n. Finding an estimate for δv_n is directly related to the eigenproblem of Q^TQ and QQ^T. These symmetric matrices have eigenvalues $\sigma_1^2,\ldots,\sigma_n^2$ which is employed in

<u>Property 3.10</u>. Let $\rho := \sigma_{n-1}-2||\delta Q||_2 > 0$.
Then $||\delta v_n||_2 \le \dfrac{||\delta Q||_2}{\sigma_{n-1}}\left[1 - 3\,\dfrac{||\delta Q||_2}{\sigma_{n-1}}\right]^{-1}$.

<u>Proof</u>. Consider the matrix $A = \begin{bmatrix} \emptyset & Q^T \\ Q & \emptyset \end{bmatrix}$ and

$A+\delta A = \begin{bmatrix} Q & (Q+\delta Q)^T \\ (Q+\delta Q) & \emptyset \end{bmatrix}$, then we can obtain the general perturbation result for symmetric matrices which says that the ratio ε of the 2-norm of the components of the perturbed vector in the subdominant and dominant direction is bounded by $\dfrac{||\delta A||_2}{\rho}$. From this it simply follows that $||\delta v_n||_2 \le \frac{\varepsilon}{1+\varepsilon}$ (cf [5]). □

We therefore conclude that $||Q^+||$ is a qualitatively appropriate measure for the conditioning. (NB $||Q^+||_2 = [\sigma_{n-1}]^{-1}$. By analogy we define the <u>sub-condition</u> number

$$(3.11) \qquad \bar{\beta} := \max ||F(t)Q^+||$$

and its estimate

(3.12) $\bar{\gamma} := ||Q^{+}||.$

§4. Applications: an algorithm for BVP on $[0,\infty)$

The previously described stability analysis can be used to investigate numerical methods for solving BVP on $[0,\infty)$ and eigenvalue problems. Here we consider as an example an algorithm that is designed to automatically compute a solution of a BVP on $[0,\infty)$ for given output values on $[0,\beta]$ say within an accuracy TOL. As we remarked earlier we have to presuppose an exponential growth of unstable solutions. For problems where they might be only polynomially growing this algorithm does not work very well and one should use other methods as described e.g. in [1, 12]. However, as we do not require any (specific) limiting behaviour of the system matrix $L(t)$, the subsequent algorithm is more generally applicable and on top of this, does the asymptotics itself.

Our method is based on marching (multiple shooting) and aims at determining the stable manifold accurately up to TOL on $[0,\beta]$, where β is some given point $< \infty$. In such a method one divides the interval under consideration into subintervals $[t_i,t_{i+1}]$ for $i= 0,\dots,N-1$, say, and computes on each subinterval a particular solution $w_i(t)$ and a fundamental solution $F_i(t)$. Essentially we employ a similar strategy as given in [21] for determining the points t_i and the initial values $F_i(t_i)$, that is we have

(4.1) $w_i(t_i) = 0, \; i = 0,\dots,N-1$

(4.2 a) $F_0(t_0)$ is orthogonal

(4.2 b) $F_{i-1}(t_i) =: F_i(t_i)U_i =: R_iU_i,$

where $F_i(t_i) = R_i$ is orthogonal and U_i upper triangular. Then there exist vectors a_i such that

(4.3) $x(t) = F_i(t)a_i + w_i(t).$

By matching (4.3) at the shooting points we then obtain

(4.4) $F_{i+1}(t_{i+1})a_{i+1} = F_{i+1}(t_{i+1})U_ia_i + w_i(t_{i+1}),$

so, denoting $g_i := R_{i+1}^T w_i(t_{i+1})$

(4.5) $$a_{i+1} = U_i a_i + g_i.$$

If we partition U_i and a_i as

(4.6) $$U_i = \begin{bmatrix} B_i & C_i \\ \emptyset & E_i \end{bmatrix}, \quad a_i = \begin{bmatrix} a_i^1 \\ a_i^2 \end{bmatrix},$$

where B_i is a k th order matrix and a_i^1 a k th order vector, we find the decoupled recursion

(4.7 a) $$a_{i+1}^2 = E_i a_i^2 + g_i^2, \quad i = 0,\ldots,N-1$$

(4.7 b) $$a_i^1 = B_i^{-1}[a_{i+1}^1 - C_i a_i^2 - g_i^2], \quad i = N-1,\ldots,0.$$

The important feature of (4.7) is that we can expect (4.7 a) to be stable in forward direction and (4.7 b) in backward direction. Indeed, fairly general initial values $F_0(t_0)$ will generate such upper triangular matrices U_i that the B_i will reflect the increments of the increasing ("unstable") modes and the E_i of the decreasing ("not unstable") modes. In [21] it is indicated how to choose $F_0(t_0)$ in the very unlikely case that $F_0(t_0) = I$ does not work satisfactorily. This decoupling not only provides a stable way to compute discrete modes via (4.7) but also a means to monitor the growth rate of the various modes (cf also [20]). In particular, at the point $\beta = t_N$ we can determine the dimension of the unstable manifold, i.e. find the integer k. This is done by simply checking the product of the diagonal elements with corresponding index of the matrices $U_0,\ldots,U_{N-1}$. Let $\mathrm{glb}(B_{N-1}\ldots B_0) =: L$, then we can anticipate an estimate for λ (cf (2.1)) of

(4.8) $$\lambda = (\ln L)/(b-a).$$

Using this λ we determine a "terminal" point γ defined by

(4.9) $$\gamma := \beta - \frac{\ln TOL}{\lambda}$$

At $t = \gamma$ the unstable solutions are then expected to have grown by a factor TOL^{-1}, compared to $t = \beta$.
We now continue our march to $t = \gamma$ introducing new shooting

points $t_{N+1},\ldots,t_M$ say. At $t = \gamma$ we check whether

$$(4.10) \qquad \|B_N^{-1}\ldots B_{M-1}^{-1}\| \le TOL$$

indeed. If not, we repeat the procedure somehow (cf. [18]). Suppose (4.10) turns out to be correct. We then compute a discrete bounded particular solution, $\{p_i^{(M)}\}_{i=0}^{M}$ say, satisfying the recursion (4.5) for $i = 0,\ldots,M-1$ and with BC

$$(4.11) \qquad p_0^2(M) = 0; \quad p_M^1(M) = 0.$$

If we assume that the recursion (4.5), formally extended to a recursion for $i \to \infty$, has some bounded particular solution, $\{q_i\}_{i=0}^{\infty}$ say, then we can define a <u>truncation error</u>

$$(4.12) \qquad t_i^1(M) := q_i^1 - p_i^1(M),$$

which apparently satisfies the homogeneous recursion

$$(4.13) \qquad t_{i+1}^1(M) = B_i t_i^1(M),$$

so

$$(4.14) \qquad t_N^1(M) = B_N^{-1}\ldots B_M^{-1} t_M^1(M).$$

Hence $\|t_N^1(M)\| = 0(TOL)$ (cf (4.10)) and a fortiori $\|t_i^1(M)\| = 0(TOL)$ for $i \le N$.

In order to complete our stable manifold we also compute a "nonincreasing" fundamental solution part $\{\Phi_i\}_{i=0}^{N}$ via (4.7) as follows: Define

$$(4.15) \qquad \Phi_0^2 := I_{n-k}, \quad \Phi_M^1 := 0$$

(Φ_i consists of n-k columns), and use (4.7 a) and (4.7 b) in the stable directions. For the same reasons as we expect $\{p_i(M)\}_{i=0}^{N}$ to be an accurate approximant up to $0(TOL)$ we expect $\{\Phi_i\}_0^N$ to be accurate up to $0(TOL)$.

The final step is the use of the BC (1.2). Since x is lying in the stable manifold there exists some (n-k) dimensional vector c such that

$$(4.16) \qquad x(t_i) = R_i[\Phi_i c + p_i(M)]$$

(for R_i see (4.2 b)).

If we substitute this in (1.2) we obtain the linear system

(4.17) $$[M_0R_0\Phi_0+M_1Q_N\Phi_N]c = b-M_0R_0p_0(M)-M_1R_Np_N(M).$$

The system (4.17) will be satisfied by some vector c up to O(TOL), if N is chosen large enough (i.e. $||B_{N-1}\cdots B_0|| \geq [TOL]^{-1}$). Therefore we solve it in the least squares sense. From the theory in section 2 we expect (n-k) singular values significantly larger than TOL and k singular values of order TOL, of course assuming well-conditioning.

Conclusion: The previously outlined algorithm determines a grid adaptively, finds an appropriate terminal value automatically and finally gives an estimate of the sub-condition number (by taking the inverse of the smallest singular value among the (n-k) largest).

§5. Some remarks about the discrete problem in §4

Theoretically we can perform the triangularizing multiple shooting strategy over the infinite interval. In the previous section we assumed that $F_0(t_0)$ was appropriately chosen. In fact it had to be such that the first k columns generate unstable modes only. This requirement, which was called *consistency* in [16] is a very weak one and bears relationships with similar (weak) requirements to make subspace iteration or the QR algorithm work. It should be realized that such a choice of $F_0(t_0)$ implies that all subsequent $F_i(t_i)$ generate only unstable modes as well; more specifically $\text{span}(F_i^1(t_i)) = \text{span}(F_0^1(t_i))$ for all i (where the superscript denotes the first k columns). From this we can deduce a nice representation of the "stable" subspace. Introduce (NB in [16], there is a misprint in (7.5))

(5.1) $$\Omega_{p,q} := -\sum_{l=p}^{q}\Big[\prod_{j=p}^{l} B_j\Big]^{-1} C_l \prod_{j=p}^{l-1} E_j$$

where we define products and sums of matrices as in

(5.2) $$\prod_{j=p}^{q} M_j = \begin{cases} M_q\cdots M_p, & q \geq p \\ I, & q < p \end{cases},$$

(5.3) $$\sum_{j=p}^{q} M_j = \begin{cases} M_p+\ldots+M_q, & q \geq p \\ 0 & , p < q \end{cases}.$$

We have

Property 5.4. Consistency and dichotomy imply that the homogeneous part of (4.5) to my has a subspace of bounded solutions represented by $\{\Psi_i^2\}$ say, where Ψ_i is an $n\times(n-k)$ matrix for all i $(\Psi_{i+1}^2 = U_i\Psi_i^2)$. If we partition Ψ_i^2 as $\begin{bmatrix}\Psi_i^{12}\\ \Psi_i^{22}\end{bmatrix}$, where Ψ_i^{22} is $(n-k)\times(n-k)$, then Ψ_i^{22} is nonsingular for all i and the direction $\Omega_i := \Psi_i^{12}[\Psi_i^{22}]^{-1}$ equals $\lim_{M\to\infty} \Omega_{i,M}$.

The result in 5.4 shows e.g. that the algorithm computes the bounded solutions asymptotically correct. For if $\psi \in \operatorname{span}(\Psi)$, then (4.7) implies

(5.5) $$\psi_i^1 = (\prod_{j=i}^{M} B_j)^{-1}\psi_M^1 + \Omega_{i,M}(\prod_{j=0}^{i-1} E_j)\psi_0^2.$$

Now an approximate solution $\phi(M)$ say (with $\phi_M^1(M) = 0$) gives

(5.6) $$\phi_i^1(M) = \Omega_{i,M}(\prod_{j=0}^{i-1} E_j)\phi_0^2$$

By identifying ϕ_0^2 with ψ_0^2 and letting $M \to \infty$ in (5.5) (using the boundedness of $\{\Phi_M^1\}$), we obtain the following "error" expression

(5.7) $$\psi_i^1-\phi_i^1(M) = (\Omega_i-\Omega_{i,M})(\ \ _{j=0}^{i-1} E_j)\psi_0^2.$$

The more precise information about the actual magnitude of this "error" follows from rewriting it as

(5.8) $$\psi_i^1-\phi_i^1(M) = (\prod_{j=i}^{M-1} B_j)^{-1}\Omega_M(\prod_{j=i}^{M-1} E_j)\psi_i^2,$$

where the factors $(\Pi B_j)^{-1}$ and (ΠE_j) show the similarity of the algorithm with inverse subspace iteration. Finally, again using 5.4 we can give an asymptotic expansion for the error, viz.

(5.9) $$\psi_i^1-\phi_i^1(M) = \sum_{l=M}^{\infty} (\sum_{j=i}^{l} B_j)^{-1}C_l(\prod_{j=i}^{l-1} E_j)\psi_i^2.$$

For the inhomogeneous case the situation is a bit more complicated. However, we can employ a discrete analogue of the expression in (2.3) to characterize some particular solution $\{q_i\}$. First we note that the discrete unstable solutions are characterized by $|\Psi_i^1\}$, with $\Psi_i^1 = \begin{bmatrix}\Psi_i^{11}\\ \emptyset^i\end{bmatrix}$, $\Psi_{i+1}^{11} = B_i\Psi_i^{11}$ (Ψ_i^{11} a k×k matrix). Here a fundamental system is given by $\{\Psi_i\}$ with

$$(5.10)\qquad \Psi_i = \begin{bmatrix}\Psi_i^{11} & \Psi_i^{12}\\ \\ \emptyset & \Psi_i^{22}\end{bmatrix} = \begin{bmatrix}(\prod_{j=0}^{i-1} B_j) & \Omega_i(\prod_{j=0}^{i-1} E_j)\\ \\ \emptyset & (\prod_{j=0}^{i-1} E_j)\end{bmatrix}.$$

If we let $P_1 = \begin{bmatrix}I_k & \emptyset\\ \emptyset & \emptyset\end{bmatrix}$, then because of the dichotomy we have the following expression for $\{q_i\}$:

$$(5.11)\qquad q_i = \sum_{\ell=0}^{i-1} \Psi_i(I-P_1)\Psi_\ell^{-1}g_\ell - \sum_{\ell=i}^{\infty} \Psi_i P_1 \Psi_\ell^{-1} g_\ell$$
$$= \sum_{\ell=0}^{i-1} \binom{\Omega_i}{I_{n-k}} (\prod_{j=\ell}^{i-1} E_j) g_\ell^2 - \sum_{\ell=i}^{\infty} \binom{I_k}{\emptyset} (\prod_{j=i}^{\ell-1} B_j)^{-1} g_\ell^1.$$

Formally writing out the expression we obtain for the particular solution $\{p_i(M)\}$ cf. (4.11))

$$(5.12)\qquad p_i^1(M) = \sum_{\ell=0}^{i-1} \Omega_{i,M}(\prod_{j=\ell}^{i-1} E_j) g_\ell^2 + \sum_{\ell=i}^{M-1} (\prod_{j=i}^{\ell-1} B_j)^{-1}\{\Omega_{\ell,M} g_\ell^2 - g_\ell^1\}.$$

Hence (cf (5.7) → (5.8))

$$(5.13)\qquad q_i^1 - p_i^1(M) = -\sum_{\ell=0}^{i-1} \sum_{s=M}^{\infty} (\prod_{j=i}^{s} B_j)^{-1} C_s (\prod_{j=\ell}^{s-1} E_j) g_\ell^2$$
$$- \sum_{M}^{\infty} (\prod_{j=i}^{\ell-1} B_j)^{-1} g_\ell^1 - \sum_{\ell=i}^{M-1} (\prod_{j=i}^{\ell-1} B_j)^{-1} \Omega_{\ell,M} g_\ell^2.$$

§6. Numerical examples

We give two examples to demonstrate the previous analysis

Example 6.1. Consider the ODE

$$(6.2)\qquad \frac{dx}{dt} = \begin{bmatrix}0 & 20\tanh^2(t)-11\\ 1 & 0\end{bmatrix} x$$

the one-dimensional Schrödinger equation, cf [4]. It has basis solutions growing like $\sim e^{3t}$ and $\sim e^{-3t}$ respectively, so $k = 1$.

First we take as BC

$$(6.3) \qquad \begin{bmatrix} 1 & 0 \\ 0 & 1 \end{bmatrix} x(0) + \begin{bmatrix} 0 & 0 \\ 1 & 0 \end{bmatrix} x(\infty) = \begin{bmatrix} 0 \\ 1 \end{bmatrix}.$$

We ask for an approximation of x on [0,4] and an accuracy TOL = 10^{-6}. The algorithm found a value for $\gamma = 5.9$ and a sub-condition number (inverse of the smallest "significant" singular value) of 1.

T	x approx	x exact	abs.error
0.00	0.105D-06	0.000D+00	0.10D-06
	0.100D+01	0.100D+01	0.25D-06
0.40	0.301D+00	0.301D+00	0.11D-07
	0.334D+00	0.334D+00	0.28D-06
0.80	0.278D+00	0.278D+00	0.17D-06
	-0.319D+00	-0.319D+00	0.24D-06
1.20	0.140D+00	0.140D+00	0.12D-06
	-0.300D+00	-0.300D+00	0.21D-06
1.60	0.538D-01	0.538D-01	0.40D-07
	-0.140D+00	-0.140D+00	0.84D-07
2.00	0.181D-01	0.181D-01	0.41D-08
	-0.510D-01	-0.510D-01	0.10D-07
2.40	0.573D-02	0.573D-02	0.23D-09
	-0.167D-01	-0.167D-01	0.53D-09
2.80	0.177D-02	0.177D-02	0.17D-09
	-0.523D-02	0.523D-02	0.57D-09
3.20	0.537D-03	0.537D-03	0.97D-10
	-0.160D-02	-0.160D-02	0.47D-09
3.60	0.163D-03	0.163D-03	0.37D-10
	-0.487D-03	-0.487D-03	0.47D-09
4.00	0.491D-04	0.491D-04	0.29D-09
	-0.147D-03	-0.147D-03	0.10D-08

Table 6.1.

The results are given in Table 6.1.
Next we take as BC

$$(6.4) \qquad \begin{bmatrix} 1 & 0 \\ 0 & 0 \end{bmatrix} x(0) + \begin{bmatrix} 0 & 0 \\ 1 & 0 \end{bmatrix} = \begin{bmatrix} 0 \\ 0 \end{bmatrix}.$$

This is in fact an eigenvalue problem formulation. If we ask for the same interval and accuracy, we get an error message that the condition number now is .367+7. However we still obtain a certain solution (now unique up to a multiplicative constant), which is just identical to the solution x in Table 6.1. The sub-condition number is 1 again.

Example 6.5. Consider the ODE (cf [21, Ex 5.1])

$$(6.6)\quad \frac{dx}{dt} = \begin{bmatrix} 1-19\cos 2t & 0 & 1+19\sin 2t \\ 0 & 19 & 0 \\ 1+19\sin 2t & & 1+19\cos 2t \end{bmatrix} x + e^{t} \begin{bmatrix} -1+19(\cos 2t-\sin 2t) \\ -18 \\ 1-19(\cos 2t+\sin 2t) \end{bmatrix}$$

(a bounded particular solution is given by $e^{t}(1,1,1)^{T}$).
The homogeneous part has solutions growing like $\sim e^{20t}$, e^{19t}, e^{-18t}, so k = 2. Note that these basis solutions rotate at "speed 1" in the (1.3) plane. We first consider the BC

$$(6.7)\quad \begin{bmatrix} 1 & 0 & 0 \\ 0 & 1 & 0 \\ 0 & 0 & 1 \end{bmatrix} x(0) = \begin{bmatrix} 1 \\ 1 \\ 1 \end{bmatrix},$$

and ask for an accuracy TOL = 10^{-6} on [0,10].
The value for γ turns out to be 11.38 and the sub-condition number 1. The numerical results are in Table 6.2.

T	y approx	y exact	abs.error
0.00	0.100D+01	0.100D+01	0.44D-07
	0.200D+01	0.200D+01	0.00D+00
	0.200D+01	0.200D+01	0.22D-14
2.00	0.135D+00	0.135D+00	0.43D-08
	0.114D+01	0.114D+01	0.14D-12
	0.135D+00	0.135D+00	0.11D-08
4.00	0.183D-01	0.183D-01	0.20D-08
	0.102D+01	0.102D+01	0.27D-12
	0.183D-01	0.183D-01	0.20D-08
6.00	0.248D-02	0.248D-02	0.11D-09
	0.100D+01	0.100D+01	0.40D-12
	0.248D-02	0.248D-02	0.27D-08
8.00	0.335D-03	0.335D-03	0.13D-08
	0.100D+01	0.100D+01	0.49D-12
	0.335D-03	0.335D-03	0.83D-09
10.00	0.454D-04	0.454D-04	0.12D-08
	0.100D+01	0.100D+01	0.55D-12
	0.454D-04	0.454D-04	0.59D-09

Table 6.2

Finally we took a similar BVP now with BC

$$(6.8)\quad \begin{bmatrix} 1 & 1 & 1 \\ 0 & 1 & 1.001 \\ 0 & 0 & 0 \end{bmatrix} x(0) = \begin{bmatrix} 3 \\ 2.001 \\ 0 \end{bmatrix}.$$

Again with TOL = 10^{-6} we now find a sub-condition number of .2001+4. From Table 6.3 it can be seen that this is manifest in the second coordinates of the approximant (compare to Table 6.2).

T	x approx	x exact	abs.error
0.00	0.100D+01	0.100D+01	0.44D-07
	0.200D+01	0.200D+01	0.44D-04
	0.200D+01	0.200D+01	·0.44D-04
2.00	0.135D+00	0.135D+00	0.43D-08
	0.114D+01	0.114D+01	0.44D-04
	0.135D+00	0.135D+00	0.11D-08
4.00	0.183D-01	0.183D-01	0.20D-08
	0.102D+01	0.102D+01	0.44D-04
	0.183D-01	0.183D+01	0.20D-08
6.00	0.248D-02	0.248D-02	0.11D-09
	0.100D+01	0.100D+01	0.44D-04
	0.248D-02	0.248D-02	0.27D-08
8.00	0.335D-03	0.335D-03	0.13D-08
	0.100D+01	0.100D+01	0.44D-04
	0.335D-03	0.335D-03	0.83D-09
10.00	0.454D-04	0.454D-04	0.12D-08
	0.100D+01	0.100D+01	0.44D-04
	0.454D-04	0.454D-04	0.59D-09

Table 6.3

References

[1] A. Bayliss, A double shooting scheme for certain unstable and singular boundary value problems, Math. Comp. 32 (1978), 61-71.

[2] C. de Boor, F. de Hoog, H.B. de Keller, The stability of one-step schemes for first-order two-point boundary value problems, SIAM J. Numer. Anal. 20 (1983), 1139-1146.

[3] W.A. Coppel, Dichotomies in Stability Theory, LNM 629, Springer-Verlag Berlin (1978).

[4] R. England, A program for the solution of boundary value problems for systems of ordinary differential equations, Culham Laboratory Report, PDN 3/73, 1976.

[5] G.H. Golub, C.F. vanLoan , Matrix computations. The John Hopkins University Press, Baltimore (1983).

[6] F.R. de Hoog, R.M.M. Mattheij, On dichotomy and well-conditioning in BVP, report 8356, Katholieke Universiteit Nijmegen (1983).

[7] F.R. de Hoog, R.M.M. Mattheij, The Role of Conditioning in Shooting Techniques, these proceedings.

[8] F.R. de Hoog, R. Weiss, An Approximation Theory for Boundary Value Problems on Infinite Intervals, Computing 24 (1980), 227-239.

[9] F.R. de Hoog, R. Weiss, On the boundary value problem for systems of ordinary differential equations with a singularity of the second kind, SIAM J. Math. Anal. 11 (1980), 41-60.

[10] H.O. Kreiss, Difference Approximtations for Boundary and Eigenvalue Problems for Ordinary Differential Equations, Math. Comp. 26 (1972), 605-624.

[11] W.F. Langford, A shooting algorithm for the best least squares solution of two-point boundary value problems, SIAM J. Numer. Anal. 14 (1977), 527-542.

[12] M. Lentini, H.B. Keller, Boundary value problems on semi-infinite intervals and their numerical solution, SIAM J. Numer. Anal. 17 (1980), 577-604.

[13] P. Markowich, A theory for the approximation of solutions of boundary value problems on infinite intervals, SIAM J. Math. Anal. 13 (1982), 484-513.

[14] J.L. Massara, J.J. Schäffer, Linear differential equations and function spaces, Academic Press, New York (1966).

[15] R.M.M. Mattheij, On approximating smooth solutions of linear singularly perturbed ODE, in: Numerical Analysis of Singular Perturbation Problems (ed. P.W. Hemker, J.J.H. Miller), Academic Press, London (1979), 457-465.

[16] R.M.M. Mattheij, Characterizations of dominant and dominated solutions of linear recursions, Numer. Math. 35 (1980), 421-442.

[17] R.M.M. Mattheij, Estimates for the errors in the solution of linear boundary value problems due to perturbations, Computing 27 (1981), 299-318.

[18] R.M.M. Mattheij, Stable computation of solutions of unstable linear initial value recursions, BIT 22 (1982), 79-93.

[19] R.M.M. Mattheij, The conditioning of linear boundary value problems, SIAM J. Numer. Anal. 19 (1982), 963-978.

[20] R.M.M. Mattheij, Estimates for the fundamental solutions of discrete BVP, J. Math. Anal. Appl. 101 (1984) 444-464.

[21] R.M.M. Mattheij, G.W.M. Staarink, An efficient algorithm for solving general linear two point BVP, SIAM J. Stat. Sci. Comp. 5 (1984), No 4.

[22] T.N. Robertson, The Linear Two-point Boundary Value Problem on an Infinite Interval, Math. Comp. 25 (1971), 475-481.

[23] R.D. Russel, A Unified View of the Numerical Solution of BVPs , these proceedings.

[24] G. Söderlind, R.M.M. Mattheij, Stability and asymptotic

estimates in nonautonomous linear differential systems, SIAM J. Math. Anal. 16 (1985), No 1.

[25] G.W. Stewart, On the perturbation of pseudo-inverses, projections and linear least squares problems, SIAM Review 19 (1977), 634-662.

[26] J.G. Verwer, A.O.H. Axelsson, Boundary value techniques for initial value problems in ordinary differential equations, manuscript (1982).

Progress in Scientific Computing, Vol. 5
Numerical Boundary Value ODEs

RICCATI TRANSFORMATIONS: WHEN AND HOW TO USE?

Paul van Loon

1. Introduction

In this paper the problem of interest is a well-conditioned n-dimensional boundary value problem (BVP):

$$\dot{x}(t) = A(t)x(t) + f(t) \qquad , t \in (0,1), \tag{1.1}$$

subject to the boundary conditions

$$B^0 x(0) + B^1 x(1) = b \tag{1.2}$$

(B^0, $B^1 \in \mathbb{R}^{n\times n}$ and $b \in \mathbb{R}^n$).
As is shown by De Hoog and Mattheij ([3]) the well-conditioning of the problem implies that the solution space of (1.1) is dichotomic.
To simplify the presentation we make the following assumptions:
on the interval (0,1) the spectrum of A can be split into a positive and a negative part. Let $\sigma(A(t)) = \{\lambda_1(t), \ldots, \lambda_n(t)\}$, then there exists a constant k, $1 \leq k < n$, such that, for all $t \in (0,1)$, $Re(\lambda_i(t)) > 0$ $(i = 1,.., k)$ and $Re(\lambda_i(t)) < 0$ $(i = k+1,.., n)$.
Moreover, assume that the invariant subspaces of A are slowly varying, such that this splitting of the spectrum implies that the differential equation (1.1) has both (rapidly) increasing and decreasing solutions (exponential dichotomy). Actually we want the eigenvalues of A to describe sufficiently well the growth behaviour of the fundamental solutions.

At one time or another any numerically stable algorithm that solves (1.1) and (1.2) has to decouple the increasing and decreasing modes (Mattheij, [6]). Here we shall look at an algorithm that is based on:
- a *Riccati transformation*, to decouple the solution space (§3)
- the *invariant imbedding* principle, to compute solution spaces of the transformed system in a numerically stable way (§2)

- *multiple shooting*, to circumvent possible instabilities and singularities (§4).

The advantage of this combination is that all initial value problems (IVPs) to be solved are well-conditioned. Especially if the BVP has solutions that vary very rapidly in magnitude (a so-called *stiff* problem), then a lot of computation time may be saved by using a stiffly stable integrator (for instance: BDF-formulas). However, we have to observe that, like other multiple shooting methods, at any restarting point a boundary layer has to be passed.
Of course, the algorithm has also some disadvantages. Since a Riccati transformation is used we have to solve a quadratic differential equation. Moreover, orthogonal transformations of (1.1) are to be executed explicitly. The most important disadvantage, however, is the fact that fast rotation of the solution spaces is disastrous for any Riccati transformation, since the transformation is based on a special parametrization of the direction of an, in general dominant, solution space. Hence, the restriction on the speed of rotation of the invariant subspaces of A is really necessary.

Finally in §5, we shall mention the kind of problems for which the method sketched above could be useful.

2. Invariant Imbedding

Before the algorithm is explained we first describe what is meant by the principle of invariant imbedding. Therefore the following assumption has to be made:
there exists a (time-dependent) Lyapunov transformation T such that for $y := T^{-1}x$ the system (1.1) becomes partially decoupled, say

$$\dot{y}(t) = \begin{pmatrix} \dot{y}_1(t) \\ \dot{y}_2(t) \end{pmatrix} \begin{matrix} \updownarrow k \\ \updownarrow n-k \end{matrix} \tag{2.1}$$

$$= \begin{matrix} k \updownarrow \\ n-k \updownarrow \end{matrix} \underbrace{\begin{bmatrix} \tilde{A}_{11}(t) & \tilde{A}_{12}(t) \\ 0 & \tilde{A}_{22}(t) \end{bmatrix}}_{k \quad\; n-k} \begin{pmatrix} y_1(t) \\ y_2(t) \end{pmatrix} + \begin{pmatrix} \tilde{f}_1(t) \\ \tilde{f}_2(t) \end{pmatrix} \quad , \; t \in (0,1).$$

Moreover, this decoupling is correctly ordered, which means that $\sigma(\tilde{A}_{11}(t)) \subset \mathbb{C}^+$ and $\sigma(\tilde{A}_{22}(t)) \subset \mathbb{C}^-$. (These conditions about the spectra may be weakened, if necessary).
As has been remarked by Mattheij ([6]) this means that the first k columns of T constitute a basis of a dominant solution space of (1.1).

Direct computation of a fundamental matrix (w.r.t. (2.1)),

$$Y(t) = \begin{bmatrix} Y_{11}(t) & Y_{12}(t) \\ 0 & Y_{22}(t) \end{bmatrix}, \text{ with } Y(0) = I_n,$$

and a particular solution

$$h(t) = \begin{pmatrix} h_1(t) \\ h_2(t) \end{pmatrix}, \text{ with } h(0) = 0,$$

would lead to:
well-conditioned IVPs for Y_{22} and h_2, but
ill-conditioned IVPs for Y_{11}, Y_{12} and h_1.
In order to obtain only well-conditoned IVPs we observe that

$$y_1(t) = Y_{11}(t)y_1(0) + Y_{12}(t)y_2(0) + h_1(t) \tag{2.2}$$

$$y_2(t) = Y_{22}(t)y_2(0) + h_2(t). \tag{2.3}$$

The earlier assumption guarantees that $Y_{11}^{-1}(t)$ exists, and so we may define $R_{11} := Y_{11}^{-1}$, $R_{12} := -Y_{11}^{-1}Y_{12}$ and $g_1 := -Y_{11}^{-1}h_1$. This leads to the relation

$$y_1(0) = R_{11}(t)y_1(t) + R_{12}(t)y_2(0) + g_1(t), \tag{2.4}$$

$t \in (0,1)$.
It can be shown that R_{11}, R_{12} and g_1 satisfy

$$\begin{aligned} \dot{R}_{11}(t) &= -R_{11}(t)\tilde{A}_{11}(t) &&, R_{11}(0) = I_k \\ \dot{R}_{12}(t) &= -R_{11}(t)\tilde{A}_{12}(t)R_{22}(t) &&, R_{12}(0) = 0 \qquad , t \in (0,1) \\ \dot{g}_1(t) &= -R_{11}(t)(\tilde{A}_{12}(t)h_2(t) + \tilde{f}_1(t)), && g_1(0) = 0. \end{aligned}$$

Observe that these IVPs are all well-conditioned.

The formulas (2.3) and (2.4) are called the invariant imbedding relations, since they express the solution y of (2.1) in terms of $y_1(t)$ and $y_2(0)$.

The main question to be answered now is the following: does there exist such a decoupling transformation T and, if it exists, how can we find it?

3. Decoupling Transformations

Since we have assumed that invariant subspaces of A are not too rapidly varying, an orthogonal basis of a dominant solution space of (1.1) at $t = 0$ is found by making a Schur-transformation of A(0). Let $U^0 = [\underbrace{U_1^0}_{k} \; \underbrace{U_2^0}_{n-k}] \in \mathbb{R}^{n\times n}$ be orthogonal and such that for $A^0 := (U^0)^T A U^0$ we have

$$A^0(0) = \begin{bmatrix} A_{11}^0(0) & A_{12}^0(0) \\ 0 & A_{22}^0(0) \end{bmatrix} \begin{matrix} \updownarrow k \\ \updownarrow n-k \end{matrix} , \tag{3.1}$$

with $\sigma(A_{11}^0(0)) = \{\lambda_1(0), \ldots, \lambda_k(0)\} \subset \mathbb{C}^+$ and $\sigma(A_{22}^0(0)) = \{\lambda_{k+1}(0), \ldots, \lambda_n(0)\} \subset \mathbb{C}^-$. The basis we look for is formed by the first k columns of U^0.

As decoupling transformation T we may choose, for instance, an orthogonal transformation or a Riccati transformation.

3.1 Orthogonal Decoupling

An orthogonal decoupling can be obtained in the following way (cf. Abramov, [1]): choose $T = [\underbrace{T_1}_{k} \;\; \underbrace{T_2}_{n-k}] \updownarrow n$ such that

$$\dot{T}_1 = (I_n - T_1 T_1^+) A T_1$$

$$\dot{T}_2 = - T_1 T_1^+ A^T T_2 \quad ,$$

subject to the initial conditon $T(0) = U^0$.

Here T_i^+ $(i = 1,2)$ means the generalized inverse of T_i $(=(T_i^T T_i)^{-1} T_i^T)$, which is used instead of T_i^T in order to keep the

numerical approximation of T orthogonal (similar remarks are made by Davey, [2]).
For this choice of T we obtain

$$\tilde{A} = \begin{bmatrix} T_1^{+}AT_1 & T_1^{+}(A^T + A)T_2 \\ 0 & T_2^{+}AT_2 \end{bmatrix} ,$$

which is, in general, correctly ordered by the initial value of T(0).

The disadvantage of this decoupling is that the IVPs are quite non-linear and moreover, the number of differential equations to be solved is almost doubled. Probably, these disadvantages are not fully compensated by the orthogonality of T, although more research in this direction seems to be worth while.

3.2 Riccati Transformation.

A more well-known way of decoupling is the Riccati transformation ([4], [6]). Normally one chooses

$$T = \begin{bmatrix} I_k & 0 \\ R_{21} & I_{n-k} \end{bmatrix} ,$$

which leads to the Riccati differential equation

$$\dot{R}_{21} = A_{21} + A_{22}R_{21} - R_{21}A_{11} - R_{21}A_{12}R_{21}$$

and

$$\tilde{A} = \begin{bmatrix} A_{11}+A_{12}R_{21} & A_{12} \\ 0 & A_{22}-R_{21}A_{12} \end{bmatrix} .$$

To obtain a correct decoupling we have to be sure that $\mathcal{R}(\begin{bmatrix} I_k \\ R_{21} \end{bmatrix})$ describes a dominant solution space. This could be forced by choosing $R_{21}(0) = U_{21}^0(U_{11}^0)^{-1}$, where U^0 is the Schur-transformation of (3.1). However, it is uncertain that $(U_{11}^0)^{-1}$ exists.

To circumvent this problem it seems better to choose

$$T = T^0 := U^0 \begin{bmatrix} I_k & 0 \\ R_{21}^0 & I_{n-k} \end{bmatrix} ,$$

since then, by the initial value $R_{21}^{0}(0) = 0$, a dominant subspace is described by the first k columns of T. This leads to the differential equation

$$\dot{R}_{21} = A_{21}^{0} + A_{22}^{0}R_{21}^{0} - R_{21}^{0}A_{11}^{0} - R_{21}^{0}A_{12}^{0}R_{21}^{0} \quad ,$$

where A^{0} is defined in (3.1).

The following questions now arise:

Is it possible to decouple the solution space by $T^{0}(t)$, for all t? And, if the answer is positive, does this always imply decoupling with a correct ordering?

Suppose it is possible. Then, from the invariant imbedding relations (2.3) and (2.4) and the boundary conditions (1.2), the solution at $t = 0$ and $t = 1$ can be computed directly. After that the solution at an interior point is explicitly expressed in terms of $(U_1^{0})^T x(1)$ and $(U_2^{0})^T x(0)$.

However, the answers to the above questions are negative. As soon as one of the solutions in the solution space, spanned by the first k columns of $T^{0}(t)$, becomes perpendicular to $\mathcal{R}(U_1^{0})$, the Riccati matrix R_{21}^{0} will blow up.

To find out how well and for how long this transformation exists and decouples correctly we look at the algebraic Riccati equation

$$0 = A_{21}^{0} + A_{22}^{0}P - PA_{11}^{0} - PA_{12}^{0}P. \tag{3.2}$$

By the special form of A^{0} and the separation of the eigenvalues one may show that, on some positive interval $(0,\bar{t})$, there exists a continuous solution of (3.2), say P_{21}^{0}, with $P_{21}^{0}(0) = 0$ (cf. Stewart, [9]). For P_{21}^{0} we know that, for all $t \in (0,\bar{t})$,

- $\mathcal{R}\left(U^{0} \begin{bmatrix} I_k \\ P_{21}^{0}(t) \end{bmatrix} \right)$ is the invariant subspace of A(t), corresponding to $(\lambda_1(t), \ldots, \lambda_k(t))$

- $\sigma(A_{11}^{0}(t) + A_{12}^{0}(t)P_{21}^{0}(t)) = \{\lambda_1(t), \ldots, \lambda_k(t)\}$

- $\sigma(A_{22}^{0}(t) - P_{21}^{0}(t)A_{12}^{0}(t)) = \{\lambda_{k+1}(t), \ldots, \lambda_n(t)\}$.

Hence, we wish R_{21}^0 to stay close to P_{21}^0.
Define, as long as R_{21}^0 and P_{21}^0 exist, $E := R_{21}^0 - P_{21}^0$. Then E satisfies

$$\dot{E} = -\dot{P}_{21}^0 + (A_{22}^0 - P_{21}^0 A_{12}^0)E - E(A_{11}^0 + A_{12}^0 P_{21}^0) - EA_{12}^0 E$$
$$E(0) = 0.$$

Therefore, successful decoupling by T^0 depends on:

1° the existence of P_{21}^0, i.e. the rotation of the corresponding invariant subspace of A

2° the way P_{21}^0 is an attractor, which depends on the separation of the eigenvalues and the magnitudes of $\dot{P}_{21}^0$ and A_{12}^0.

At a certain moment, however, even in the smoothly varying case, the Riccati matrix may blow up or decouple incorrectly. By that time we have to make some kind of restart.

4. Multiple Shooting

Assume restarts are necessary at the points $t = t_i$ ($i = 1,.., m-1$) and let $t_m = 1$. On each interval (t_i, t_{i+1}) the decoupling transformation T^i is now chosen as

$$T^i = U^i \begin{bmatrix} I_k & 0 \\ R_{21}^i & I_{n-k} \end{bmatrix},$$

where U^i has to make this decoupling correctly ordered. For U^i ($i = 1,.., m$) we have in principle two possibilities:

1 - U^i the Schur-transformation for which the matrix $A^i(t_i) := (U^i)^T A(t_i) U^i$ is quasi-uppertriangular and correctly ordered.

This has the advantage that the decoupling is certainly done correctly and that the total number of restarts, m, is nearly minimized. The disadvantage, however, is that for each restart a Schur-transformation has to be computed. Moreover, after $t = 1$ is reached, a (maybe large) multiple shooting system has to be solved.

2 - Choose $U^i = [\underbrace{U_1^i}_{k} \ \underbrace{U_2^i}_{n-k}]$ orthogonal, such that at $t = t_i$ the columns

of $U_1^{\,i}$ form an orthogonal basis of the solution space which is described by $\mathcal{R}(U^{i-1}\begin{bmatrix} I_k \\ R_{21}^{\,i-1}(t_i) \end{bmatrix})$.
This same solution space is in the interval (t_i, t_{i+1}) given with regard to the new basis U^i.
This choice of U^i has the advantage that during the integration an LU-decomposition of the final multiple shooting matrix is obtained (cf. stabilized march). Therefore, we only have to solve one n×n system to determine $(U_2^{\,0})^T x(0)$ and $(U_1^{\,m})^T x(1)$. Other values of the solution can then be obtained by simple matrix-vector multiplications.
This choice of U^i may lead to a somewhat larger number of restarts.

Probably the safest strategy is the following: try the second possibility of U^i. Accept this choice if $A^i(t_i) := (U^i)^T A(t_i) U^i$ is well-ordered and otherwise take for U^i the Schur-transformation of $A(t_i)$. In general, rejection will hardly arise.

If no restarts using a Schur-transformation are necessary, then the final solution is computed similar to the case of no restart at all. If some Schur-transformations are necessary, say ℓ, then a multiple shooting system of size $(\ell+1)n$ has to be solved.

Remark:
Of course, if the boundary conditions are separated one could also choose U^0 such that

$$B^0 U^0 = \begin{bmatrix} 0 & 0 \\ 0 & * \end{bmatrix} \begin{matrix} \updownarrow k \\ \updownarrow n-k \end{matrix} .$$

In that case some reduction is achieved, since the corresponding Y_{22} and R_{12} need not to be computed. However, this classical choice ([8]) may lead to an incorrectly ordered decoupling, as is shown by the next example:

$$\dot{x}(t) = \begin{bmatrix} -10 & 0 \\ 20 & 10 \end{bmatrix} x(t) \qquad , \; t \in (0,1)$$

subject to $x_2(0) = 0$ and $x_2(1) = 1$.
This is a well-conditoned BVP. Corresponding to the boundary condition at $t = 0$ we have $U^0 = I$, which implies that $\tilde{A}_{11}^{\,0} = -10$ and $\tilde{A}_{22}^{\,0} = 10$.

5. Applications

Although the ideas are not fully implemented yet, we may describe the kind of problems for which the method will be useful.
The directions of fundamental solutions are parametrized in a very restricted way. Hence, good performance of the algorithm may only be expected if fundamental solutions are slowly rotating. Solutions with rapid change in magnitude make the differential equations harder to solve, but not dramatically. This is the main reason why variants of this method are used for singular perturbation problems ([7], [10]).
As an example we have solved problem 3 of [4]:

$$\dddot{y}(t) = w\ddot{y}(t) + \dot{y}(t) - wy(t), \qquad t \in (0, T),$$

subject to the boundary conditions

$$y(0) = e^{-wT} + 0.1\,(1 + e^{-T})$$
$$y(T) = 1 + 0.1\,(1 + e^{-T})$$
$$\dot{y}(T) = w + 0.1\,(1 - e^{-T}).$$

Exact solution: $y(t) = e^{w(t-T)} + 0.1(e^{t-T} + e^{-t})$.

The stiffness of this problem is determined by the parameters w and T. Most of the tested codes in [4] were already in trouble for relatively small values of w and T ($w = 20$ and $T = O(1)$).
We have tried our code for $w = 20$ and $T = 100$ and with an accuracy of 10^{-6}. As is shown in the second column of table 1 the code behaves exactly as we expected it to do: in each boundary layer it uses small stepsizes, but once this layer has been passed the stepsize is extremely increased. Since the Riccati matrix will never blow up the total performance is mainly determined by the number of points where output is demanded (at each point an internal layer is created). This effect is shown by the third column of table 1, since the code is arranged in such a way that the amount of work is almost completely determined by the number of function evaluations.

length of interval	stepsize at end of interval	number of function evaluations
1	0.0555	112
2	0.129	124
4	0.221	139
8	0.375	154
16	0.496	176
32	5	193
100	20 (!)	220

table 1. (w = 20)

The stiffness of the problem has only a small effect on the performance. Even for w = 20,000 and with 6 equidistributed output points we did not meet any problem (see table 2).

w	stepsize at end of subinterval (length: 20)	number of function evaluation per subinterval
20	2.89	184
200	1.99	203
2000	0.753	214
20000	0.901	220

table 2. (T = 100)

Since in the constant coefficients case the Riccati matrix R_{21} is always identically zero, this example illustrates how useful the Schur-transformation and the invariant imbedding principle may be.

Secondly we have chosen an example where the invariant subspaces of A are rotating (cf. Example 9.1 of [6]):

$$x(t) = \begin{bmatrix} 1 + 19 \cos 2\omega t & 0 & -\omega + 19 \sin 2\omega t \\ & 19 & 0 \\ \omega + 19 \sin 2\omega t & 0 & 1 - 19 \cos 2\omega t \end{bmatrix} + f(t), \quad t \in (0, \pi),$$

subject to the boundary condition

$$x(0) + x(\pi) = (1 + e^{\pi}, \omega(1 + e^{-\pi}), 1 + e^{\pi})^T.$$

Now f is chosen such that the solution becomes $x(t) = (e^t, \omega e^{-t}, e^t)^T$. The example is constructed such that on the first subinterval we have $R_{21}(t) = \tan(\omega t)$ and $\sigma(A(t)) = \{19, 1 \pm \sqrt{345}\}$. Hence, depending on the value of ω and our restart strategy, some restarts will be necessary. If a restart is made as soon as one of the elements of $R_{21}(t)$ becomes in absolute value larger than a prescribed constant p, the following results for $\omega = 4$ are obtained (accuracy: 10^{-6}).

p	number of restarts	per subinterval	
		number of integration steps	number of function evaluations
1	15	50 - 60	80 - 95
10	8	90 - 100	170 - 240

table 3.

In full agreement with our expectation, the number of subintervals is halved by choosing p = 10 instead of p = 1. For both values of p the amount of work is almost identical, but for values of p larger than 10 it increases very fast. Moreover, the decoupling will in that case become incorrectly ordered. Hence, it is advised to choose p not too large.

Another area of interest is BVPs having a singularity of the first kind. For these problems one may show that the boundary condition

$$\lim_{t \downarrow 0} x(t) \text{ exists}$$

can be rewritten such that the number of differential equations reduces. Moreover, all solutions we look for will be analytic, which gives us the possibility, using power series, to move away from the singularity ([5]).

In the near future more research into different aspects of the algorithm has to be done. For instance, for stiff BVPs it must be possible to circumvent the creation of internal layers at restarting points.
Another area of interest involves the algebraic problems to be solved when one is trying to get some optimal implementation. Probably these aspects will be worked out before the next workshop on BVPs.

Paul van Loon
Computing Centre
Eindhoven University of Technology
P.O. Box 513
5600 MB EINDHOVEN
The Netherlands

References

[1] Abramov, A.A. : 'On the transfer of the condition of boundedness for some systems of ordinary linear differential equations', USSR Comp. Math. and Math. Phys., 1 (4), (1961), p. 875-881.

[2] Davey, A. : 'An automatic orthonormalization method for solving stiff BVPs', Journ. Comp. Phys., 51, (1983), p. 343-356.

[3] De Hoog, F.R. and R.M.M. Mattheij : 'On dichotomy and well-conditioning in BVP', report 8355, Dept. of Mathematics, Catholic University Nijmegen.

[4] Lentini, M., M.R. Osborne and R.D. Russell: 'The close relationship between methods for solving two-point BVP's, to appear in SAIM J. Num. Anal.

[5] Van Loon, P.M. : 'Reducing a singular linear two-point BVP to a regular one by means of Riccati transformations', report 83-WSK-03, Eindhoven University of Technology.

[6] Mattheij, R.M.M. : 'Decoupling and stability of BVP algorithms', report 8314, Dept. of Mathematics, Catholic University Nijmegen (to appear in SIAM Review).

[7] Mattheij, R.M.M. and R.E. O'Malley, Jr. : 'On solving BVPs for multi-scale systems using asymptotic approximations and multiple shooting', report 8353, Dept. of Mathematics, Catholic University Nijmegen (to appear in BIT).

[8] Meyer, G.H. : 'Initial value methods for boundary value problems', Academic Press, New York, 1973.

[9] Stewart, G.W. : 'Error bounds for approximate invariant subspaces of closed linear operators', SIAM J. Num. Anal., 8, (1971), p. 796-808.

[10] Weiss, R. : 'An analysis of the box and trapezoidal schemes for linear singularly perturbed BVPs', Math. Comp., 42, (1984), p. 41-67.

Progress in Scientific Computing, Vol. 5
Numerical Boundary Value ODEs

DISCRETIZATIONS WITH DICHOTOMIC STABILITY FOR TWO-POINT BOUNDARY VALUE PROBLEMS

Roland England, Robert M.M. Mattheij

1. Introduction

For a two-point boundary value problem to be well conditioned, the system of ordinary differential equations must necessarily possess a dichotomic set of fundamental solutions [7], with decaying modes controlled by initial conditions, and growing modes controlled by terminal conditions [10]. It was shown in [3] that it is important for a discretization of such a problem to preserve the dichotomy property, and the implications of this stability criterion were examined in a number of particular cases. Some simple difference schemes were examined for second order ordinary differential equations, and also various discretizations for first order systems, including those obtained by piecewise collocation and implicit Runge-Kutta type formulae. The last two examples were of multistep schemes, of such a form that they could be used in a sequential stepping mode, as would be done in shooting methods, or more generally in multiple shooting.

In this paper, two families of such schemes are presented, both having desirable dichotomic stability characteristics. The motivating stability analysis resembles that used for stiff initial value problems, concentrating on the linear test equation, $dy/dt = \lambda y$, but differing in that great attention is paid to values of λ with large positive real parts. Using the one-step member of one of the families, some preliminary results are also presented, showing the feasibility of forward integration in such a situation, by using appropriately large step sizes where the particular solution is sufficiently smooth.

2. Dichotomic Stability

A general linear scheme for first order differential equations may consist of one or a number of linear homogeneous formulae relating approximations to the solution and its derivatives, where the derivatives are obtained by substitution from the differential equation. The usual linear stability analysis is carried out by applying the scheme to the standard linear test equation, $dy/dt = \lambda y$, for arbitrary complex values of λ, and a fixed step size h. Where more than one formula is involved, intermediate approximations are eliminated between the formulae to obtain a single k-step recurrence relation of the form:

$$Q_0(h\lambda)y_{i+1} + Q_1(h\lambda)y_i + \dots + Q_k(h\lambda)y_{i-k+1} = 0 \quad ; \qquad (1)$$

where $Q_j(z)$, $j = 0,1,\dots,k$, are polynomials in z, of degree s which is normally the number of stages (individual formulae), or the number of derivatives involved in a single formula. Such a recurrence relation has a general solution of the form:

$$y_i = \sum_{j=1}^{k} c_j R_j(h\lambda)^i \quad ; \qquad (2)$$

where $R_j(z)$, $j = 1,2,\dots,k$, are the roots of the characteristic or stability polynomial:

$$Q(z,R) = Q_0(z)R^k + Q_1(z)R^{k-1} + \dots + Q_k(z) \quad ; \qquad (3)$$

and c_j, $j = 1,2,\dots,k$ are arbitrary constants.

In [3] a number of new stability properties were defined in the context of multiple shooting. Where a multistep scheme is used to solve initial value problems, in a sequential stepping mode, additional starting values must be generated by some special starting procedure. Since these constitute additional initial conditions, the corresponding spurious modes of the recurrence relation (1) should be of decaying type, in order that the initial values should control them. However, the principal mode of the discretization should be of decaying type if $h\lambda$ has negative real part, but of growing type if $h\lambda$ has positive real part, so as to preserve the dichotomy occurring in the fundamental solutions of the differential equation. Let the roots $R_j(z)$ of the characteristic polynomial (3) be ordered in such a way that $R_1(z)$ is the principal root, satisfying $R_1(z) - e^z = O(z^{p+1})$ as $z \to 0$, where p is the order of consistency of the discretization. The above stability requirements imply that:

$$\begin{aligned} &\text{(i)} && |R_1(z)| \le 1 && \text{if } \mathcal{Re}(z) \le 0 \quad ; \\ &\text{(ii)} && |R_1(z)| \ge 1 && \text{if } \mathcal{Re}(z) \ge 0 \quad ; \\ &\text{(iii)} && |R_j(z)| \le 1 && \text{for } j = 2,3,\ldots,k \quad . \end{aligned} \tag{4}$$

It appears that these conditions cannot be satisfied for all complex values of z, except by a subset of linearly symmetric one-step schemes, for which $k = 1$, and $R_1(z) = 1/R_1(-z)$ is a rational function whose poles all have positive real part. For some purposes, these schemes require excessive computational effort, and it is desirable to examine multistep schemes which satisfy conditions (4) for values of z in a large part of the complex plane, although they cannot do so everywhere.

For values of z with negative real part, conditions (4) are exactly those required for A-stability, as defined in [1]. In that context, the requirement of A-stability also puts severe restrictions upon the available schemes, and for many circumstances, it has been relaxed to that of $A(\alpha)$-stability [12], for which it is required that conditions (4) be satisfied in some region

$$R^- := \{z \in \mathbb{C} \; | \mathcal{Re}(z)\sin\alpha \le -|\mathcal{Im}(z)|\cos\alpha\} \quad . \tag{5}$$

In a similar way, it is useful to consider schemes which satisfy the conditions (4) in some region $R^- \cup R^+$ where

$$R^+ := \{z \in \mathbb{C} \; | \mathcal{Re}(z)\sin\alpha \ge |\mathcal{Im}(z)|\cos\alpha\} \quad . \tag{6}$$

3. Dichotomically Stable Multistep Formulae

Although it is not necessary for dichotomic stability, it seems useful to require that $R_1(z)$ remain bounded in some region of the form R^+. By use of the theory of order stars [11], this implies that the positive real axis of the complex plane z lies within a white finger of the order star, so that the formula is relatively stable according to the definition:

$$|R_j(z)| \leq |e^z| \quad \text{for} \quad j = 1,2,\ldots,k \quad ; \tag{7}$$

in some region of the form R^+. Also $R_1(z)$ must remain real and greater than unity along the positive real axis, so that its limit as $z \to \infty$ will be $+1$, which will mean that the principal mode of the discretization will be smooth rather than oscillatory.

In terms of the characteristic polynomial, the condition that $R_1(z)$ have no pole on the real axis implies that $Q_0(z)$ must have no real roots. Also, it was shown in [12] that for $A(\alpha)$-stability a scheme must be implicit, and thus $Q_0(z)$ may not be a mere constant. The simplest form for $Q_0(z)$ is therefore a quadratic polynomial, and to avoid the introduction of unnecessary stages or derivatives, the characteristic polynomial would take the form:

$$Q(z,R) = -\sum_{j=0}^{k} (\alpha_j+\beta_j z+\gamma_j z^2)R^{k-j} \quad ; \tag{8}$$

where, for normalization, one may arbitrarily fix $\alpha_0 = -1$.

For a general multistage scheme, the order conditions are quite complicated, but in terms of the characteristic polynomial they reduce to $R_1(z) - e^z = O(z^{p+1})$ as $z \to 0$, which imposes $p + 1$ conditions on the $3(k+1)$ coefficients α_j, β_j, γ_j. As already mentioned, it is also useful to require that $R_1(z) \to +1$ as $z \to \infty$, which imposes a further condition, while those given by (4) are in general inequalities. In order to ensure that conditions (4) are satisfied in a large neighbourhood of the origin $z = 0$, and also of $z = \infty$, it is convenient to have

$$R_j(0) = 0 \quad \text{and} \quad R_j(z) \to 0 \quad \text{as} \quad z \to \infty, \text{ for } j = 2,3,\ldots,k \quad . \tag{9}$$

This in turn implies that:

$$\alpha_1 = 1 \quad , \quad \alpha_2 = \alpha_3 = \ldots = \alpha_k = 0 \quad , \tag{10}$$

and

$$\gamma_1 = -\gamma_0 \quad , \quad \gamma_2 = \gamma_3 = \ldots = \gamma_k = 0 \quad ; \tag{11}$$

which reduces the form (8) of the characteristic polynomial to:

$$Q(z,R) = R^k - R^{k-1} - z \sum_{j=0}^{k} \beta_j R^{k-j} - z^2\gamma(R^k - R^{k-1}) \quad ; \qquad (12)$$

where the remaining $k + 2$ coefficients, β_j $(j = 0,1,\ldots,k)$ and γ, may be chosen to satisfy the order conditions with $p = k + 2$.

For $p \leq 11$, the resulting schemes do in fact satisfy conditions (4) in regions of the form $R^- \cup R^+$ with values of α which decrease as p increases, but for $p > 11$ the schemes are no longer $A(\alpha)$-stable.

4. Two Specific Families

The linear stability properties of a scheme may be studied purely in terms of the coefficients of its characteristic polynomial, but those coefficients do not uniquely determine a specific scheme. However, given a characteristic polynomial (3), a corresponding linear k-step s-derivative scheme is given by:

$$Q_0(hD)y_{i+1} + Q_1(hD)y_i + \ldots + Q_k(hD)y_{i-k+1} = 0 \quad ; \qquad (13)$$

where D is the differentiation operator d/dt with respect to the independent variable t. Thus, for a characteristic polynomial of the form (8), a general linear second derivative multistep scheme is given by:

$$\sum_{j=0}^{k} (\alpha_j y_{i-j+1} + \beta_j h\dot{y}_{i-j+1} + \gamma_j h^2 \ddot{y}_{i-j+1}) = 0 \quad , \qquad (14)$$

the form from which Enright began his development of second derivative schemes for stiff initial value problems [5]. In particular, dichotomically stable second derivative schemes, with characteristic polynomial (12) are given by:

$$y_{i+1} = y_i + h \sum_{j=0}^{k} \beta_j \dot{y}_{i-j+1} + h^2\gamma(\ddot{y}_{i+1} - \ddot{y}_i) \quad ; \qquad (15)$$

where the local truncation error is $O(h^{p+1})$ as $h \to 0$, with $p = k + 2$.

In [2], a family of two-stage multistep schemes was developed with the same characteristic polynomials as Enright's second derivative schemes, and therefore of use for stiff initial value problems. In the same way, a family of dichotomically stable two-stage multistep schemes may be set up, having the characteristic polynomial (12), but not requiring the evaluation of second derivatives. One formula is an interpolation formula for the solution at an off-step point:

$$y_{i+\theta} = \sum_{j=0}^{k} a_j y_{i-j+1} + h\alpha(\dot{y}_{i+1}-\dot{y}_i) \quad ; \tag{16}$$

where the coefficients a_j $(j = 0,1,\ldots,k)$ and α are functions of θ which are chosen so that the local truncation error is $O(h^p)$ as $h \to 0$. The other formula is a quadrature rule making use of the derivative at the off-step point for a fixed value of θ:

$$y_{i+1} = y_i + h \sum_{j=0}^{k} b_j \dot{y}_{i-j+1} + h\beta\dot{y}_{i+\theta} \quad ; \tag{17}$$

where the coefficients b_j $(j = 0,1,\ldots,k)$ and β are functions of θ chosen so that the local truncation error is $O(h^{p+1})$ as $h \to 0$. This is the same quadrature rule as that required in [2], in the context of stiff initial value problems.

Since both (15) and (17) must be identically satisfied whenever $y(t)$ is a polynomial of degree not exceeding p, then by subtracting one from the other:

$$h\beta\dot{y}_{i+\theta} = h \sum_{j=0}^{k} (\beta_j-b_j)\dot{y}_{i-j+1} + h^2\gamma(\ddot{y}_{i+1}-\ddot{y}_i) \tag{18}$$

whenever $\dot{y}(t)$ is a polynomial of degree less than p. With $p = k + 2$, the coefficients a_j $(j = 0,1,\ldots,k)$ and α of (16) are unique functions of θ, and by comparison with (18) they must satisfy the following relations:

$$\beta a_j = \beta_j - b_j \quad (j = 0,1,\ldots,k) \quad ;$$
$$\beta\alpha = \gamma \quad . \tag{19}$$

To use these Hybrid Implicit Dichotomic schemes (16), (17), it is necessary to choose specific values of θ for each order. As in [2], this may be done by setting b_k to zero, which permits the saving of a little storage space, as well as giving rise to values of θ between 0 and 1 (off-step points within the interval of interest). In [9], an alternative criterion is proposed for selecting θ , based on the local truncation error for the linear test equation.

5. Fourth Order Schemes and Test Problem

Examining schemes with $p = 4$, it is found that the coefficient $\beta_2 = 0$, so that for this case (1) is only a one-step recurrence, and (15) is a symmetric one-step second derivative scheme, which satisfies conditions (4) for all complex values of z :

$$y_{i+1} = y_i + \tfrac{1}{2}h(\dot{y}_{i+1}+\dot{y}_i) - \frac{1}{12}h^2(\ddot{y}_{i+1}-\ddot{y}_i) \quad . \tag{20}$$

In the corresponding two-stage case, if θ is chosen so that $b_2 = 0$, then it is also found that $a_2 = 0$, and (16), (17) form a symmetric two-stage implicit Runge-Kutta scheme:

$$\begin{aligned} y_{i+\frac{1}{2}} &= \tfrac{1}{2}(y_{i+1}+y_i) - \tfrac{1}{8}h(\dot{y}_{i+1}-\dot{y}_i) \quad ; \\ y_{i+1} &= y_i + h[\tfrac{1}{6}\dot{y}_i + \tfrac{2}{3}\dot{y}_{i+\frac{1}{2}} + \tfrac{1}{6}\dot{y}_{i+1}] \quad ; \end{aligned} \tag{21}$$

which equally satisfies conditions (4) for all complex values of z . This scheme, or other equivalent formulations of it, have appeared in the literature a number of times, since as long ago as 1957 ([8] page 206, block method of Clippinger and Dimsdale). It may also be obtained by collocation at Lobatto quadrature points, and appeared again in a slightly different form in [6]. It may be put into the usual Runge-Kutta form by substituting the second formula into the first.

It is this one-step fourth order scheme (21) which has been implemented into a variable step integrator, and used to obtain preliminary results. The step control mechanism had to be specially designed to cope with modes of the type $dy/dt = \lambda y$ where λ has large positive real parts. This design, and the success thereof, are reported in [4]. The essential point is that appropriately large step sizes are used where the particular solution being followed is sufficiently smooth, and that this causes the unstable modes to grow only slowly, and not to blow up within each shooting interval.

The integrator has been applied to the initial value problem:

$$\frac{dY}{dt} = \begin{bmatrix} 9 & -12 & -5 & 0 \\ -12 & -9 & 0 & -5 \\ 15 & 0 & 9 & -12 \\ 0 & 15 & -12 & -9 \end{bmatrix} Y + F(t) \quad ; \qquad (22)$$

where $F(t)$ is such that $Y = [0 \;\; 0 \;\; e^{20t/\mu} \;\; 2e^{-20t/\mu}]^T$ is an exact solution. The initial conditions were chosen to correspond to the exact solution, or perturbations thereof, and the problem was solved for various values of μ and of the tolerance for the integrator. It should be noted that the complementary functions of (22) have growth behaviour $e^{(\pm 15 \pm 5i\sqrt{3})t}$, and therefore, as compared to the smooth solution, they are faster the larger the value of μ.

The results shown in table 1 show the number of steps taken, the final value of t reached, and the global error observed at that point. When the integration was stopped owing to a dramatic reduction in step size (blow-up of the unstable modes), this has been noted, but it only occurs when excessive precision is requested, or the initial perturbation exceeds the tolerance. In all other cases, the global error observed may be explained as the sum of the initial perturbation (virtually unchanged) and the global truncation error. Thus, the results demonstrate the feasibility of forward integration with large step sizes, even in the presence of strongly unstable modes.

6. References

[1] G.G. Dahlquist: A special stability problem for linear multistep methods, BIT, 3 (1963), pp.27-43.

[2] R. England: Some hybrid implicit stiffly stable methods for ordinary differential equations, Numerical Analysis Proceedings, Cocoyoc, Mexico (ed. J.P. Hennart), Lecture Notes in Mathematics 909, Springer (1982), pp.147-158.

[3] R. England, R.M.M. Mattheij: Boundary value problems and dichotomic stability, K.U. Nijmegen Dept. of Maths. report 8356 (1983).

[4] R. England, R.M.M. Mattheij: Sequential step control for integration of two-point boundary value problems, to appear in Numerical Analysis Proceedings, Guanajuato, Mexico (1984), Springer Lecture Notes in Mathematics.

[5] W.H. Enright: Second derivative multistep methods for stiff ordinary differential equations, SIAM J. Numer. Anal., 11 (1974), pp.321-331.

[6] J.P. Hennart, R. England: A comparison between several piecewise continuous one step integration techniques, Working Papers for the 1979 SIGNUM Meeting on Numerical Ordinary Differential Equations, Champaign, Illinois (ed. R.D. Skeel), University of Illinois at Urbana-Champaign, Dept. of Computer Science report 963 (1979), pp.33.1-33.4.

[7] F.R. de Hoog, R.M.M. Mattheij: On dichotomy and well-conditioning in BVP, K.U. Nijmegen Dept. of Maths. report 8355 (1983).

[8] K.S. Kunz: Numerical Analysis, 1st edition, McGraw-Hill (1957).

[9] M. Lautsch: An implicit off-step method for the integration of stiff differential equations, Computing, 31 (1983), pp.177-183.

[10] R.M.M. Mattheij: The conditioning of linear boundary value problems, SIAM J. Numer. Anal., 19 (1982), pp.963-978.

[11] G. Wanner, E. Hairer, S.P. Nørsett: Order stars and stability theorems, BIT, 18 (1978), pp.475-489.

[12] O.B. Widlund: A note on unconditionally stable linear multistep methods, BIT, 7 (1967), pp.65-70.

Roland England,
IIMAS - UNAM,
Apdo. Postal 20-726,
01000 México D.F.,
Mexico.

Robert M.M. Mattheij,
Mathematisch Instituut,
Katholieke Universiteit,
6525 ED Nijmegen,
The Netherlands.

With partial support from the Netherlands Organization for the Advancement of Pure Research (ZWO).

FURTHER FIGURES AND TABLES

Fig. 1 shows the boundaries of absolute stability, where $|R_j(z)| = 1$, for the dichotomically stable multistep formulae with characteristic polynomial (12). As p increases there is a region of instability which grows to the left of the imaginary axis. Figs. 2 and 3 show the order stars for $p = 4$ and 5 , and it may be seen that condition (7) is satisfied in some region of the form R^+ with relatively small values of α , which are in fact equal to $90^o/(p+1)$ for $p > 4$. The order stars also display greater detail of the stabilities for values of z close to the origin, including for example, the instability along the imaginary axis for $p = 5$. Table 2 shows the coefficients of the characteristic polynomial (12) for the dichotomically stable multistep formulae, and also the values of α for which the schemes are $A(\alpha)$-stable.

TABLE 1

Results for test problem:

Exact Initial Values at $t = 0$

no. of steps/final t / global error

Tolerance per step	$\mu = 10^4$	10^5	10^6	10^7
10^{-4}	$14/5 \times 10^2$ 6×10^{-9}	$17/5 \times 10^3$ 6×10^{-10}	$20/5 \times 10^4$ 5×10^{-11}	$45/5 \times 10^5$ 7×10^{-13}
10^{-7}	$81/5 \times 10^2$ 4×10^{-9}	$84/5 \times 10^3$ 2×10^{-11}	$121/5 \times 10^4$ 1×10^{-12}	$521/5 \times 10^5$ 8×10^{-15}
10^{-10}	$9/5.0$ 2×10^{-11} Collapse $h \gtrsim 10^{-2}$	$63/2.9 \times 10^3$ 2×10^{-9} Collapse $h \gtrsim 10^{-3}$	$208/2.6 \times 10^3$ 5×10^{-10} Collapse $h \gtrsim 10^{-2}$	$280/3.4 \times 10^3$ 2×10^{-8} Collapse $h \gtrsim 10^{-4}$

TABLE 1 (Continued)

Initial Values Perturbed by approximately 10^{-9}				
Tolerance per step	$\mu = 10^4$	10^5	10^6	10^7
10^{-4}	$14/5 \times 10^2$ 1×10^{-9}	$17/5 \times 10^3$ 6×10^{-10}	$20/5 \times 10^4$ 1×10^{-9}	$45/5 \times 10^5$ 1×10^{-9}
10^{-7}	$81/5 \times 10^2$ 2×10^{-7}	$84/5 \times 10^3$ 1×10^{-9}	$133/5 \times 10^4$ 1×10^{-9}	$1332/5 \times 10^5$ 1×10^{-9}
10^{-10}	$866/1.25$ 9×10^{-3}	0/0.0 Collapse $h \approx 10^{-2}$	6/67.4 Collapse $h \approx 10^{-2}$	$16/2 \times 10^2$ 5×10^{-9} Collapse $h \approx 10^{-3}$

TABLE 1 (Continued)

Initial Values Perturbed by 5×10^{-6}				
Tolerance per step	$\mu = 10^4$	10^5	10^6	10^7
10^{-4}	$14/5 \times 10^2$ 5×10^{-6}	$17/5 \times 10^3$ 5×10^{-6}	$20/5 \times 10^4$ 5×10^{-6}	$42/5 \times 10^5$ 5×10^{-6}
10^{-7}	$5/20.0$ Collapse $h \gtrsim 10^{-3}$	$240/4.8 \times 10^3$ Collapse $h \gtrsim 10^{-5}$		
10^{-10}	$0/0.0$ Collapse $h \gtrsim 10^{-3}$	$0/0.0$ Collapse $h \gtrsim 10^{-3}$		

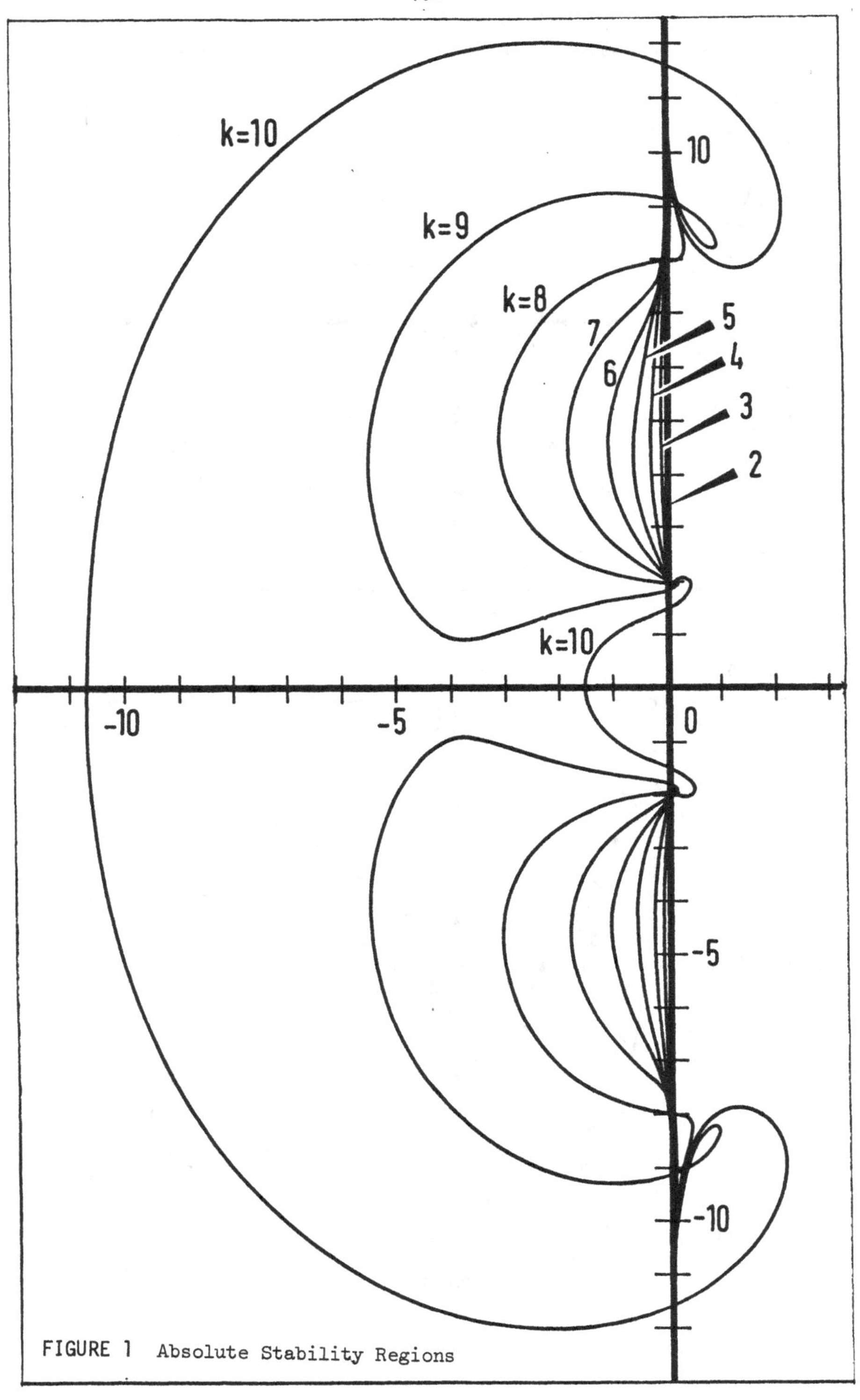

FIGURE 1 Absolute Stability Regions

ORDER STAR OF 4TH ORDER SCHEME

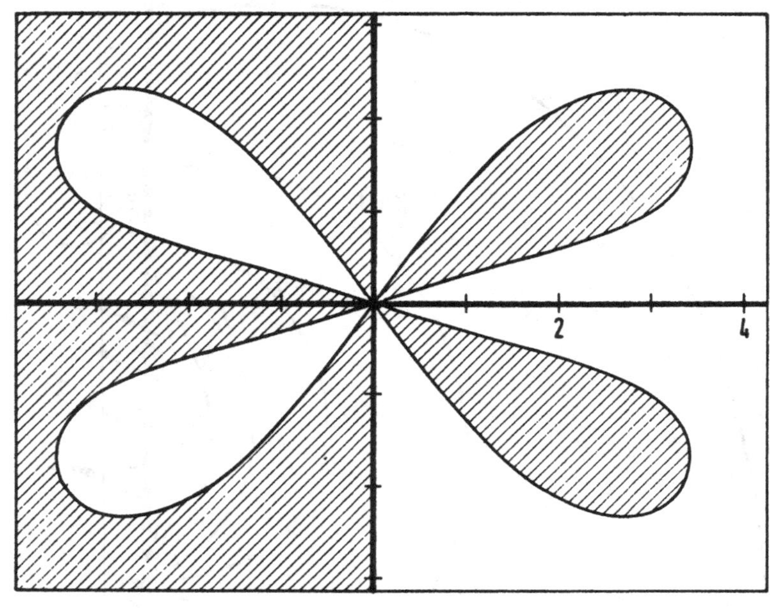

FIGURE 2

ORDER STAR OF 5TH ORDER SCHEME

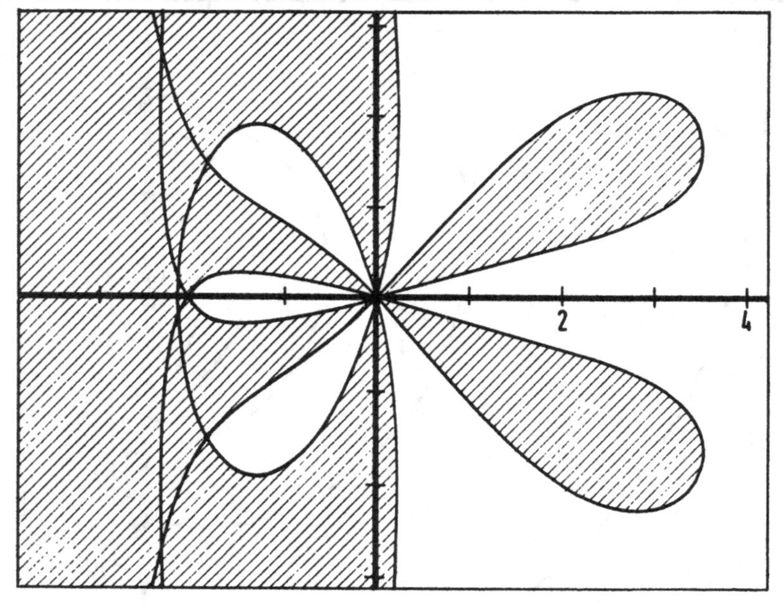

FIGURE 3

TABLE 2

Coefficients of characteristic polynomials and value of α in degrees

k	2	3	4	5	6	7	8	9	10
p	4	5	6	7	8	9	10	11	12
α	90	88.3	85.3	81.0	74.8	65.5	50.4	12.9	
γ	$\frac{-1}{12}$	$\frac{-19}{240}$	$\frac{-3}{40}$	$\frac{-863}{12096}$	$\frac{-275}{4032}$	$\frac{-33953}{518400}$	$\frac{-8183}{129600}$	$\frac{-3250433}{53222400}$	$\frac{-4671}{78848}$
β_0	$\frac{1}{2}$	$\frac{79}{160}$	$\frac{35}{72}$	$\frac{69455}{145152}$	$\frac{16289}{34560}$	$\frac{4817897}{10368000}$	$\frac{4160257}{9072000}$	$\frac{6750165673}{14902272000}$	$\frac{218386751}{487710720}$
β_1	$\frac{1}{2}$	$\frac{247}{480}$	$\frac{77}{144}$	$\frac{404213}{725760}$	$\frac{20039}{34560}$	$\frac{43743001}{72576000}$	$\frac{11345701}{18144000}$	$\frac{28946608421}{44706816000}$	$\frac{1631753477}{2438553600}$
β_2	0	$\frac{-1}{96}$	$\frac{-7}{240}$	$\frac{-473}{8640}$	$\frac{-193}{2240}$	$\frac{-32989}{268800}$	$\frac{-49577}{302400}$	$\frac{-304001}{1451520}$	$\frac{-20669129}{79833600}$
β_3		$\frac{1}{480}$	$\frac{7}{720}$	$\frac{4637}{181440}$	$\frac{149}{2880}$	$\frac{392381}{4354560}$	$\frac{96833}{680400}$	$\frac{16754303}{79833600}$	$\frac{4696471}{15966720}$
β_4			$\frac{-1}{720}$	$\frac{-5207}{725760}$	$\frac{-5219}{241920}$	$\frac{-433609}{8709120}$	$\frac{-213103}{2177280}$	$\frac{-18378821}{106444800}$	$\frac{-9994091}{35481600}$
β_5				$\frac{221}{241920}$	$\frac{7}{1280}$	$\frac{152219}{8064000}$	$\frac{298637}{6048000}$	$\frac{173595227}{1596672000}$	$\frac{67891267}{319334400}$
β_6					$\frac{-19}{30240}$	$\frac{-471389}{108864000}$	$\frac{-57733}{3402000}$	$\frac{-6623411}{133056000}$	$\frac{-1078613}{8870400}$
β_7						$\frac{9829}{21772800}$	$\frac{9623}{2721600}$	$\frac{1737823}{111767040}$	$\frac{28288223}{558835200}$
β_8							$\frac{-407}{1209600}$	$\frac{-26449691}{8941363200}$	$\frac{-25825123}{1788272640}$

TABLE 2
continued

k	9	10
p	11	12
β_9	$\frac{330157}{1277337600}$	$\frac{9668683}{3832012800}$
β_{10}		$\frac{-24377}{119750400}$

Progress in Scientific Computing, Vol. 5
Numerical Boundary Value ODEs

IMPROVING THE PERFORMANCE OF NUMERICAL METHODS FOR TWO POINT BOUNDARY VALUE PROBLEMS*

by

W.H. Enright
Department of Computer Science
University of Toronto

Abstract

We report on an ongoing investigation into the performance of numerical methods for two point boundary value problems. We outline how methods based on multiple shooting, collocation and other local discretizations can share a common structure. The identification of this common structure permits us to analyse how various components of a method interact and also permits us to consider the assembly of a collection of modular routines which will eventually form the basis for a software environment for solving two point boundary value problems.

The initial stage of our investigation has involved the implementation and analysis of a family of multiple shooting methods as well as a family of collocation/Runge-Kutta methods. We have analysed the performance of these methods on a class of singular perturbation problems. The numerical conditioning of both families of methods on such problems and the convergence requirements of the corresponding iteration schemes (used to solve the discretized problem) has been investigated. Appropriate modifications to these methods which permit the effective solution of such problems will be discussed. We will also identify a subfamily of the Runge-Kutta methods that are particularly effective.

---

*This research was supported by the National Science and Engineering Research Council of Canada.

1. Introduction

To understand and implement effective general purpose numerical methods for two point boundary value problems now involves the use of an appropriate discretization of the differential equation; a reliable adaptive mesh selection strategy; a robust nonlinear equation solver and special purpose linear algebra routines. The performance of a method can be adversely affected by any of these components and it is therefore instructive to investigate how they interact before attempting to implement a new approach or improve the performance of an existing approach.

As a model problem we will consider a system:

$$y' = f(t,y), \quad t\varepsilon[a,b], \quad y\varepsilon R^n \tag{1}$$

with separated boundary conditions:

$$Ay(a) = y_a, \quad By(b) = y_b$$

where A is an $(n-k)\times n$ matrix and B a $k\times n$ matrix, both of full rank.

Most methods designed for this problem can be extended in a natural and straightforward way to handle more general problems such as: non-separable boundary conditions, nonlinear interior constraints (multi-point problems), parameter continuation, eigenvalues and overdetermined systems (parameter fitting); but much can be learned by considering the performance of a specific method on this problem.

In contrast to the situation for initial value problems, there are relatively few general purpose methods available for (1). Those that are available usually fall into one of three groups: multiple shooting (eg. BVPSOL of Deuflhard and Bader (1982)), collocation (eg. COLSYS of Ascher, Christiensen and Russel (1979)), and finite difference with deferred correction (eg. PASVA3 of Lentini and Pereyra (1977)). These methods, although widely used and appreciated, can fail unexpectedly and can be difficult to use effectively without a detailed knowledge of the underlying algorithm.

To better understand how difficulties can arise and to gain insights into how some of these may be overcome we will introduce a high level overview of a typical two point boundary value method. We will then interpret multiple shooting methods and collocation/Runge-Kutta methods from this point of view. In each case we will identify where improvements in efficiency are possible, consider some specific numerical difficulties which can arise and suggest new approaches to help overcome these difficulties.

2. Overview of a method

Inevitably methods for (1) are based on a modified Newton iteration to find an approximation determined by a discrete solution, Y, satisfying $g(Y)=0$. Generally Y will be composed of discrete approximations $\{y_i\}_{i=0}^{N}$ to the true solution of (1), $y(t)$, evaluated on a mesh $a=t_o<t_1\dots<t_N=b$. The 'residual function', $g(Y)$, will vary in form and complexity but usually has the special structure:

$$g(Y) = \begin{bmatrix} Ay_o-y_a \\ \phi_1(y_o,y_1) \\ \vdots \\ \phi_N(y_{N-1},y_N) \\ By_N-y_b \end{bmatrix} . \tag{2}$$

The Jacobian matrix required on iteration ℓ of the Newton iteration, $\left.\frac{\partial g}{\partial Y}\right|_{Y^\ell}$, will frequently be approximated in some way $(W^\ell \approx \left.\frac{\partial g}{\partial Y}\right|_{Y^\ell})$ and it is this approximation that motivates the term '<u>modified Newton</u>' iteration.

A typical two point boundary value method can then be interpreted as a two level iteration scheme with the structure illustrated in Fig. 1. Note that, because of (2), the matrix W^ℓ will have a 'staircase' or near 'block bi-diagonal' form:

$$W^\ell = \begin{bmatrix} A & 0 & 0 & \dots & 0 \\ L_1 & R_1 & 0 & \dots & 0 \\ 0 & L_2 & R_2 & \dots & 0 \\ \vdots & \vdots & & & \\ 0 & 0 & \dots & L_N & R_N \\ 0 & 0 & \dots & 0 & B \end{bmatrix} , \tag{3}$$

where $L_i \approx \left.\frac{\partial \phi_i}{\partial y_{i-1}}\right|_{Y^{\ell-1}}$ and $R_i \approx \left.\frac{\partial \phi_i}{\partial y_i}\right|_{Y^{\ell-1}}$ are both $n\times n$ matrices. Considerable savings in storage and time for step 6 of Fig. 1 are possible if one uses linear algebra packages which exploit this special structure. A recent survey (Fourer (1984)) discusses what packages are available.

repeat r=1,2,... *until* error exit *or* satisfactory results

1. Determine a mesh $a=t_o^r<t_1^r...<t_N^r=b$
   - and corresponding residual $g_r(Y)$
2. Determine an initial approximation Y^o
3. *repeat* $\ell=1,2,...$NMAX *or until* convergence
   4. Evaluate the residual $g_r(Y^{\ell-1})$
   5. Determine the iteration matrix

$$W^\ell \simeq \left.\frac{\partial g_r}{\partial Y}\right|_{Y^{\ell-1}}$$

   6. Solve for the Newton correction δ^ℓ:

$$W^\ell \delta^\ell = -g(Y^{\ell-1}).$$

   7. Set $Y^\ell = Y^{\ell-1}+\delta^\ell$

end repeat

end repeat

Fig. 1 : Structure of a Typical Method

For our tests we have used the Fortran packages of Diaz, Fairweather and Keast (1981). For most methods the dominant costs are associated with steps 4. and 5. of Fig. 1., where the residual function is evaluated and the corresponding iteration matrix, W^ℓ is determined.

3. Multiple shooting methods

To interpret multiple shooting from the viewpoint of Fig. 1 we first introduce, for a given mesh $a=t_o<t_1...<t_N=b$, a local initial value problem over each subinterval:

$$z_i' = f(t,z_i),\ z(t_{i-1}) = y_{i-1},\ \text{over } [t_{i-1},\ t_i]. \tag{4}$$

Multiple shooting is then based on replacing (1) with the equivalent larger, but computationally more robust problem: $g(Y)=0$, where $y=[y_o y_1 ... y_N]^T$ and the ϕ_i (of (2)) are determined by continuity:

$$g(Y) = \begin{bmatrix} Ay_o - y_b \\ z_1(t_1) - y_1 \\ \vdots \\ z_N(t_N) - y_N \\ By_N - y_b \end{bmatrix}.$$

An evaluation of g(Y) thus requires a sequence of initial value integrations to determine $z_i(t_i)$, i=1,2,...N and standard initial value software can be used. To evaluate the iteration matrix W^ℓ we observe that $R_i=-I$ while $L_i = \frac{\partial}{\partial y_{i-1}}[z_i(t)]\big|_{t=t_i}$ can be determined by letting $L_i(t) = \frac{\partial}{\partial y_{i-1}}[z(t)]$ and integrating (with standard initial value software) the matrix initial value problem:

$$\frac{d}{dt}[L_i(t)] = \frac{\partial f}{\partial y}\bigg|_{y=z_i(t)} L_i(t), \ L_i(t_{i-1}) = I, \qquad (5)$$
$$\text{over } [t_{i-1}, t_i].$$

It should be noted that an alternative approximation can be generated using numerical differences at a comparable cost.

With the implementation outlined above we see that each inner iteration of multiple shooting (steps 4-7 of Fig. 1) requires the solution of a sequence of initial value problems of size n^2+n. An interesting feature of this method is that if the inner iteration converges, it converges quadratically to the true solution ($Y^\ell \to \{y(t_i)\}_{i=0}^N$) and the only significance of the outer iteration is to select a mesh and initial guess Y^o so that the inner iteration will converge. Most of the numerical difficulties experienced with multiple shooting can be attributed to the scaling and conditioning of the linear equations arising in the Newton iteration or to the failure of the inner iteration to converge.

Numerical conditioning of the Newton system (step 6. of Fig. 1) can be a serious problem since the blocks L_i, for well conditioned boundary value problems, must possess both growing and decaying solution components (see, for example, England and Matheij (1983)). If the length of a mesh interval, $(t_i - t_{i-1})$ is large relative to the time constants present one can either experience overflow in the computation of L_i or the columns can become numerically dependent. Restricting the maximum mesh interval can control the size of $\|L_i\|$ and also avoid numerically singular blocks, but the number of mesh intervals and the resulting increase in storage may become intolerable. In a later section we will consider the use of stable inaccurate initial value integrations as one approach to resolving this difficulty.

Even when the Newton system is well-conditioned, the convergence of the inner iteration can be very sensitive to the accuracy of the initial guess. Two standard approaches to help cope with this difficulty are the use of continuation (see Keller (1976)) and the use of a more robust nonlinear equation solver such as Gauss-Newton or damped-Newton. One way to view continuation is that for the early outer iterations of Fig. 1, $r=1,2...k$ the differential equation (and hence the residual function g_r) changes smoothly from a well behaved problem defining g_1 to the more difficult original problem (1) defining g_k. A related approach which we are investigating is to fix the differential equation but use some other discretization in the early iterations to determine more accurate starting values for the multiple shooting discretization. For example, g_1 could correspond to a low order collocation scheme and g_k to a multiple shooting scheme, both on the same mesh, with intermediate g_j's corresponding to higher order collocation schemes.

4. Collocation/implicit Runge-Kutta methods

A collocation method is based on approximating $y(t)$ by a piecewise polynomial defined on a specified mesh. On each interval $(t_{i-1}t_i)$ the polynomial $p_i(t)$ is defined, in terms of the parameters $\alpha_1,\alpha_2...\alpha_s$, by continuity conditions and s collocation conditions:

$$f(t_{i-1}+\alpha_j(t_i-t_{i-1}),\ p_i(t_{i-1}+\alpha_j(t_i-t_{i-1}))) = p_i'(t_{i-1}+\alpha_j(t_i-t_{i-1})), \quad \text{for } j=1,2,\ldots s. \tag{6}$$

To interpret this method from the viewpoint of Fig. 1 we introduce a basis set for the space of piecewise polynomials and associate the discrete solution Y with the representation of $p_i(x)$ (in this basis) for $i=1,2...N$. The size of Y is then at least snN and the blocks L_i,R_i of W^ℓ are $sn\times sn$ matrices.

A class of methods closely related to collocation methods are those based on Runge-Kutta discretizations of the differential equation where a Runge-Kutta formula is applied at t_{i-1} and g is defined in terms of the amount that the discrete approximation y_i fails to satisfy the Runge-Kutta formula. That is, if z_i is the exact solution to the Runge-Kutta equation:

$$z_i = y_i+(t_i-t_{i-1})\sum_{r=1}^{s} w_r k_r \ , \tag{7a}$$

where the k_r's are the solutions of:

$$k_r = f(t_{i-1}+\alpha_r(t_i-t_{i-1}),\ y_i+(t_i-t_{i-1})\sum_{j=1}^{s} \beta_{rj}k_j),\ r=1,2\ldots s, \tag{7b}$$

then

$$g(Y) = \begin{bmatrix} Ay_o-y_a \\ z_1-y_1 \\ \vdots \\ z_N-y_N \\ By_N-y_b \end{bmatrix} .$$

Since z_i is an approximation to $z_i(t_i)$ it is clear that, on a fine mesh, a Runge-Kutta method will be closely related to multiple shooting. The blocks R_i and L_i are $n\times n$ matrices and Y is defined in the usual way as $\{y_i\}_{i=0}^{N}$. It is well known (Wright (1970), Weiss (1974)) that for any collocation method (determined by the mesh and $\alpha_1,\alpha_2\ldots\alpha_s$) there exists a Runge-Kutta method which is equivalent in the sense that if they both converge the discrete solution generated by the Runge-Kutta method will agree with the piecewise polynomial, of the collocation method, evaluated at the meshpoints.

There can be considerable computational advantage gained by implementing a collocation method as a Runge-Kutta method or in deriving and implementing other Runge-Kutta methods that do not correspond to collocation methods. Before we explore this in more detail we will review some notation and results from the literature on Runge-Kutta formulas and general stability properties that are desirable in formulas used in boundary value discretizations.

An s-stage Runge-Kutta formula, (7a) and (7b), can be represented by the tableau:

$$\begin{array}{c|cccc} \alpha_1 & \beta_{11} & \beta_{12} & \cdots & \beta_{1s} \\ \alpha_2 & \beta_{21} & \beta_{22} & \cdots & \beta_{2s} \\ \vdots & \vdots & \vdots & & \\ \alpha_s & \beta_{s1} & \beta_{s2} & \cdots & \beta_{ss} \\ \hline & \omega_1 & \omega_2 & \cdots & \omega_s \end{array} \tag{8}$$

When applied to the linear scalar problem $y'=\lambda y$, $h=(t_i-t_{i-1})$, the Runge-Kutta solution satisfies:

$$z_i = \frac{P_s(h\lambda)}{Q_s(h\lambda)} y_{i-1} , \tag{9}$$

where $P_s(\omega)$ and $Q_s(\omega)$ are polynomials in ω of degree bounded by s and the rational function, $\frac{P_s(\omega)}{Q_s(\omega)}$ is called the stability function associated with the Runge-Kutta formula. If the coefficients of the tableau (8) satisfy $\beta_{rj}=0$ for $j\geq r$, then the formula is an <u>explicit</u> formula and, in this case, g and W will be inexpensive to compute (step 4. and 5. of Fig. 1). The major disadvantage of explicit formulas is that the corresponding stability function degenerates to a polynomial (as the degree of $Q_s(\omega)$ is zero) and this gives poor representation of decaying components. As we have already noted, decaying components are necessarily present in well-conditioned boundary value problems and therefore this property of explicit Runge-Kutta formulas can lead to numerical difficulty on some well-conditioned problems (an extremely fine mesh may be required to obtain any solution at all).

There has recently been considerable discussion concerning what forms of stability functions are best suited for methods for boundary value problems and there appears to be two schools of thought. The first approach (see for example Kreiss and Kreiss (1982), Ringhofer (1982)) is to select formulas that are 'fitted to the problem' with a uniform or almost uniform mesh. In this approach two formulas are used, one with the degree of $P_s(\omega)$ = degree of $Q_s(\omega)+1$ and the other with the degree of $P_s(\omega)$ = degree of $Q_s(\omega)-1$. The former formula is well-suited to represent increasing components and the latter well-suited to represent decaying components.

On each mesh the sign(s) of the real part of the eigenvalues of $\frac{\partial f}{\partial y}$ are determined and the appropriate formula chosen. While well suited to scalar problems there is some difficulty in applying this approach to systems of equations as the problem must then be 'uncoupled' on each mesh interval and different formulas applied to different components. The numerical conditioning of the resulting Newton system (step 6. of Fig. 1.) can be sensitive to the correct matching of formula to each component.

The second approach (see for example Ascher and Weiss (1983)) is to fix the formula with a non-uniform mesh 'fitted' to the problem. With this approach, the formula should have a stability function with degree

$P_s(\omega)$ = degree $Q_s(\omega)$. The resulting Newton systems are generally well-conditioned and robust although the cost and accuracy obtained can be very sensitive to the correct choice of mesh.

Whichever approach one adopts it is always possible to choose Runge-Kutta formulas with the right form of stability function but the formulas will generally be implicit and relatively expensive to implement. The evaluation of the residual will involve a sequence of N non-linear systems, each of order sn (to solve for the k_r's of (7b) on each mesh interval) while the blocks L_i will require $O(n^3s^2)$ operations to compute. In particular the general collocation method characterised in (6), corresponds to an equivalent Runge-Kutta method with the tableau:

$$\begin{array}{c|cccc} \alpha_1 & \int_0^{\alpha_1} \ell_1(t)dt & \int_0^{\alpha_1} \ell_2(t)dt & \cdots & \int_0^{\alpha_1} \ell_s(t)dt \\ \alpha_2 & \int_0^{\alpha_2} \ell_1(t)dt & \int_0^{\alpha_2} \ell_2(t)dt & \cdots & \int_0^{\alpha_2} \ell_s(t)dt \\ \cdot & & & & \\ \cdot & & & & \\ \cdot & & & & \\ \alpha_s & \int_0^{\alpha_s} \ell_1(t)dt & \int_0^{\alpha_s} \ell_2(t)dt & \cdots & \int_0^{\alpha_s} \ell_s(t)dt \\ \hline & \int_0^{1} \ell_1(t)dt & \int_0^{1} \ell_2(t)dt & \cdots & \int_0^{1} \ell_s(t)dt \end{array} \qquad (10)$$

where $\ell_j(t) = \prod_{\substack{r=1 \\ r\neq j}}^{s} \frac{(t-\alpha_r)}{(\alpha_j-\alpha_r)}$ for j=1,2,...s. This will be an implicit formula and the resulting Runge-Kutta implementation will have comparable operation counts associated with each inner iteration (steps 4-7) of Fig. 1, although there will be a significant storage reduction.

The obvious question that arises is whether there exist Runge-Kutta methods with suitable stability functions that do not have the large operation counts associated with each inner iteration of a general implicit formula. Muir (1984) has investigated such a class of Runge-Kutta formulas which are closely related to the class of mono-implicit formulas developed by Cash and Moore (1980) and van Bokhoven (1980). A related set of Runge-Kutta formulas has been proposed independently by Gupta (1983). Muir has analysed the stability and efficiency of these formulas in the context of boundary value problems. One interesting subclass of the formulas he considers is those characterised by tableaus of the form:

$$
\begin{array}{c|ccccc}
0 & 0 & 0 & \cdots & 0 \\
1 & \omega_1 & \omega_2 & & \omega_s \\
\alpha_3 & \beta_{31} & \beta_{32} & 0.. & 0 \\
\cdot & \cdot & \cdot & & \\
\cdot & \cdot & \cdot & & \\
\cdot & \cdot & \cdot & & \\
\alpha_s & \beta_{s1} & \beta_{s2} & ..\beta_{ss-1} & 0 \\
\hline
 & \omega_1 & \omega_2 & \cdots & \omega_s
\end{array}
\qquad (11)
$$

From the first two rows of the tableau we observe that $k_1=f(t_{i-1},y_{i-1})$ and $k_2=f(t_i,z_i)$ while the remaining k_j's are 'implicit' only in terms of k_2. At convergence we have that $z_i=y_i$ and therefore k_2 will equal $f(t_i y_i)$. One can therefore consider an equivalent discretization:

$$z_i = y_{i-1}+h\omega_1 k_1+h\omega_2 f(t_i,y_i)+\sum_{r=3}^{s}\omega_r\hat{k}_r , \qquad (12a)$$

where $h = (t_i-t_{i-1})$ and

$$\hat{k}_r = f(t_i+\alpha_r h,y_{i-1}+h\beta_{r1}k_1+h\beta_{r2}f(t_i,y_i)+h\sum_{j=3}^{r-1}\beta_{rj}\hat{k}_j) \qquad (12b)$$

$$\text{for } r = 3,4...s.$$

Note that, at convergence, the $\hat{k}_r$ of (12b) will agree with the k_r of (11) and hence the discrete solutions will be identical. On the other hand, in the implementation of (12a)-(12b), the evaluation of $g(Y)$ and the determination of W^ℓ (steps 4. and 5. of Fig. 1) will be explicit inexpensive calculations. It is also worth noting that the value $f(t_i,y_i)$ required on mesh i is k_1 on the $(i+1)^{st}$ mesh interval and only (s-1) evaluations of the differential equation are required per mesh interval. Muir has shown that formulas of the form (11) exist that are of order s and with the corresponding stability function satisfying degree $P_s(\omega)$ = degree $Q_s(\omega)$ or degree $P_s(\omega)$ = degree $Q_s(\omega)\pm1$ for $s\leq5$. It must be acknowledged that while the cost per iteration is reduced using this approach, it is not clear whether the convergence properties of the inner iteration will be different. In our limited numerical experience we have not noticed any difference in either the likelihood of achieving convergence (from the same Y^o) or in the rate of convergence.

A common difficulty shared by collocation/Runge-Kutta methods is that it can be very difficult to obtain convergence on an initial mesh. Typically mesh selection schemes require a converged solution on an initial mesh before they can begin to match the mesh to the particular problem. In this case uniform mesh refinement is performed until a converged solution is obtained. For some problem, in particular for some singular perturbation problems, this initial uniform mesh refinement can lead to very large systems which exhaust available storage and cause the method to fail. Although this behaviour is similar in some respects to that observed with multiple shooting on the same class of problems, it is much easier to resolve as a more appropriate initial mesh, graded in the boundary layers, will avoid the difficulty. In the case of multiple shooting the rapid growth of $L_i(t)$ and $z_i(t)$ will severely constrain the mesh size (t_i-t_{i-1}) on the whole range and more effort is required to avoid this difficulty. We have developed a hybrid technique to help cope with this type of difficulty and have found that it can prove very effective on some problems. It can be thought of as either a modification to a Runge-Kutta method where unknowns are removed or as a modification to a multiple shooting method where inexact (but stable) initial value integrations are used. In this latter context Sharp and Enright (1984) present results to show the improvements in performance that can be realized with multiple shooting methods that employ this technique.

This hybrid technique can be motivated by letting Y be defined by $\{y_i\}_{i=0}^{N}$ as before and letting z_i be the approximation to the local initial value problem, (4) obtained starting from $(t_{i-1}y_{i-1})$ and taking 2^k steps of a Runge-Kutta formula with stepsize $(t_i-t_{i-1})/2^k$. With

$$g(Y) = \begin{bmatrix} Ay_o-y_b \\ z_1-y_1 \\ \vdots \\ z_N-y_N \\ By_N-y_b \end{bmatrix}$$

we then have a method which with k=0 reduces to a standard Runge-Kutta method. Also for any $k\geq 1$, if the scheme converges, it converges to a discrete solution that is identical to that which would have been obtained by applying the Runge-Kutta method on the fine mesh. Furthermore it is clear that k can vary from one mesh interval to another and that as k increases the scheme reduces to multiple shooting since $z_i \to z(t_i)$.

The evaluation of g for this scheme (step 4. of Fig. 1) requires the application of $2^k N$ steps of the underlying Runge-Kutta formula and if formulas of the form (11) are chosen this involves $2^k N$ systems of equations, each of size n. To evaluate W one can either use divided differences or inexact integrations of the variational equations. Note that the size of the blocks of W will be $n \times n$; the number of unknowns will be determined by the coarse mesh and the accuracy will be determined by the fine mesh.

One advantage of interpreting existing methods from the viewpoint of Fig. 1 is that similarities between different methods are revealed. These similarities often suggest possibilities for constructing promising hybrid techniques. In this paper we have identified a few of these possibilities. We have implemented the techniques we have discussed and are currently carrying out numerical experiments to quantify the improvements that can be realized.

References

U. Ascher, J. Christiansen, and R.D. Russell (1979), "A collocation solver for mixed order systems of boundary value problems", Math. Comp. 33, 659-679.

U. Ascher and R. Weiss (1983), "Collocation for singular perturbation problems I: first order systems with constant coefficients", SIAM J. Number. Anal. 20, 537-557.

W.M.G. van Bokhoven (1980), "Efficient higher order implicit one-step methods for integration of stiff differential equations", BIT 20, 34-43.

J.R. Cash and D.R. Moore (1980), "A high order method for the numerical solution of two-point boundary value problems", BIT 20, 44-52.

P. Deuflhard and G. Bader (1982), "Multiple shooting techniques revisited", Preprint No. 163, Institut für Angewandte Mathematik, University of Heidelberg.

J.C. Diaz, G. Fairweather, and P. Keast (1981), "COLROW and ARCECO: Fortran packages for solving certain block diagonal linear systems by modified alternate row and column elimination", ACM Trans. on Math. Soft. 9, pp. 358-377.

R. England and R.M.M. Mattheij (1983), "Boundary value problems and dichotomic stability", Report 8356, Dept. of Mathematics, University of Nijmegen.

R. Fourer (1984), "Staircase matrices and systems", SIAM Review 26, 1-70.

S. Gupta (1984), "An adaptive boundary value Runge-Kutta solver for first order boundary value problems", this proceedings.

J.B. Keller (1976), "Numerical Solution of two point boundary value problems", Regional Conference Series in Applied Mathematics, No. 24, SIAM Philadelphia.

B. Kreiss and H.O. Kreiss (1981), "Numerical methods for singular perturbation problems", SIAM J. Numer. Anal. 18, 262-276.

M. Lentini and V. Pereyra (1977), "An adaptive finite difference solver for non-linear two-point boundary value problems with mild boundary layers", SIAM J. Numer. Anal. 14, 91-111.

P. Muir (1984), "Runge-Kutta methods for two point boundary value problems", Ph.D. Thesis, Dept. of Computer Science, University of Toronto.

C. Ringhofer (1982), "On collocation schemes for quasilinear singularly perturbed boundary value problems", Technical Summary Report, Mathematics Research Center, University of Wisconsin-Madison.

P.W. Sharp and W.H. Enright (1984), "On solving linear singular perturbation two-point boundary value problems using multiple shooting", manuscript.

R. Weiss (1974), "The application of implicit Runge-Kutta and collocation methods to boundary value problems", Math. Comp. 28, 449-464.

K. Wright (1970), "Some relationships between implicit Runge-Kutta, collocation and Lanczos τ methods and their stability properties", BIT 20, 217-227.

Progress in Scientific Computing, Vol. 5
Numerical Boundary Value ODEs

Reducing the Number of Variational Equations in the Implementation of Multiple Shooting*

Fred T. Krogh
(Jet Propulsion Laboratory,
California Institute of Technology, Pasadena, CA)

J.P. Keener
(University of Utah, Salt Lake City, UT)

Wayne H. Enright
(University of Toronto, Toronto, Canada)

Abstract

The standard method of multiple shooting for a system of n first order differential equations, with k unknown initial conditions requires the integration of k sets of variational equations on the first shot, and n sets of variational equations on every shot thereafter. This paper describes a variant of multiple shooting that requires the solution of k sets of variational equations on every shot. The technique applies to both linear and nonlinear boundary value problems. Techniques to deal with difficulties unique to the solution of nonlinear problems are suggested.

*This research was carried out at the Jet Propulsion Laboratory, California Institute of Technology, under contract with the National Aeronautics & Space Administration.

1. Introduction

In comparison with finite difference methods, simple shooting methods for the solution of boundary value problems in ordinary differential equations have the advantages that: sophisticated software for the solution of initial value problems can be used, an initial guess for the solution is only required at a single point (no advantage if one can easily make equally good guesses throughout the interval), and less storage is required for problems which require many mesh points to get a sufficiently accurate solution. Unfortunately, the initial value problems which must be solved when using a simple shooting method frequently have rapidly growing solutions which quickly dominate the desired solution. Even with the true initial values, a numerical solution may overflow. Multiple shooting permits the solution of such problems at the expense of requiring a little additional storage, and guesses for the initial values at a few additional points.

The idea of shooting from both ends of the interval and matching in the middle is an old one that is still of use for some problems. Apparently the first paper to generalize this idea to take more than two shots is that by Morrison et al. [1]. More recently multiple shooting has been popularized by Keller [2], [3], Osborne [4], Deuflhard [5] and others. The basic approach in multiple shooting simply involves starting shots from various points in the interval, and imposing a continuity condition wherever one shot meets another. It is known, (e.g. [6], [7]), that multiple shooting can be useful in avoiding conditioning problems.

The form of multiple shooting in [1] has the advantage of introducing fewer unknowns than more recently suggested algorithms. In the interior of the interval one gets two shots, one forward and one backward, from each set of unknown initial values. Of course there may be problems in integrating both forwards and backwards (either due to a different rate of growth in the spurious solutions, or because the problem requires taking data sequentially from a magnetic tape) and automatic selection of points where shots are to be started is more difficult. The more usual approach is to integrate from one breakpoint to the next, always in the same direction, where a breakpoint is the place where one shot ends and another begins. It is this latter form of multiple shooting that we are using.

Our interest is in a general purpose algorithm which provides the user with automatic selection of the breakpoints if he so desires, and which as much as possible permits the user to take advantage of any particular structure a problem may have. In this paper we consider the very frequently occuring case that n-k of the solution values are known at one end of the interval, thus leaving only k initial values to be determined. Each unknown initial value requires the integration of one set of variational equations, each set of variational equations involving the same number of equations as the original system. Although only k sets of variational equations are required on the first shot, the usual implementation of multiple shooting requires the integration of n sets on every succeeding shot.

We present a technique which requires the integration of k sets of variational equations on every shot. In fact we present two versions of our technique, one of which is equivalent to standard multiple shooting in that the only discrepancy in the results would be due to errors in the solution of the associated initial value problems and linear systems of equations. One way to view our technique is that it is one way of extending the techniques and principles of stabilised march ([8],[9],[10],[11] and [7]) to nonlinear problems as well as to problems with eigenvalues. Including eigenvalues, $\underset{\sim}{\lambda}$, explicitly in our problem specification does complicate the algebra but it leads to a more effective technique than would be possible if one were to instead solve additional differential equation of the form $\underset{\sim}{\lambda}'=0$. Readers not interested in eigenvalue problems or who are content to work with a larger system of ODEs can safely ignore all references to $\underset{\sim}{\lambda}$.

2. Multiple Shooting

Consider the boundary value problem

(2.1) $$y' = f(t,y,\lambda), \quad t\varepsilon(a,b)$$

(2.2) $$Ay(a) = y_a, \quad By(b) = y_b,$$

where $y\varepsilon\mathbb{R}^n$, $\lambda\varepsilon\mathbb{R}^m$ $(m\geq 0)$, $f:[a,b]\times\mathbb{R}^n\times\mathbb{R}^m\to\mathbb{R}^n$, and the matrices $A:\mathbb{R}^n\to\mathbb{R}^{n-k}$ $(k>0)$ and $B:\mathbb{R}^n\to\mathbb{R}^{m+k}$ have full row rank.

Let $a=x_o<x_1<\ldots<x_{p-1}<x_p=b$, and let $y_i(t,\lambda)$ be a solution of eq. (2.1) with the initial condition $y_i(x_i,\lambda)=s_i$, $i=0,1,\ldots,p-1$. Multiple shooting involves the solution of the system

(2.3) $$\begin{cases} g_o = As_o - y_a = 0 \\ g_i = y_{i-1}(x_i,\lambda)-s_i = 0, \quad i = 1,2,\ldots,p-1, \\ g_p = By_{p-1}(x_p,\lambda)-y_b = 0 \end{cases}$$

for $s_o,s_1,\ldots,s_{p-1}$, and λ. The points $x_1,x_2,\ldots,x_{p-1}$ where the continuity conditions $g_1,\ldots,g_{p-1}=0$ are imposed, are called breakpoints. Let $Y_i(t,\lambda)=\partial y_i/\partial s_i$ and $Y_i^{(\lambda)}(t,\lambda)= \partial y_i/\partial\lambda$. These matrices satisfy the matrix differential equations

(2.4) $$\frac{dY_i}{dt} = \frac{\partial f}{\partial y}Y_i, \quad Y_i(x_i,\lambda) = I,$$

(2.5) $$\frac{dY_i^{(\lambda)}}{dt} = \frac{\partial f}{\partial y}Y_i^{(\lambda)}+\frac{\partial f}{\partial\lambda}, \quad Y_i^{(\lambda)}(x_i,\lambda) = 0,$$

and Newton's iteration for solving eq. (2.3) is given by

(2.6) $$\begin{bmatrix} A & & & 0 \\ Y_o(x_1,\lambda)-I & & & Y_o^{(\lambda)}(x_1,\lambda) \\ & \ddots & & \vdots \\ & & Y_{p-2}(x_{p-1},\lambda)-I & Y_{p-2}^{(\lambda)}(x_{p-1},\lambda) \\ & & BY_{p-1}(x_p,\lambda) & BY_{p-1}^{(\lambda)}(x_p,\lambda) \end{bmatrix} \begin{bmatrix} \Delta s \\ \Delta s_1 \\ \vdots \\ \Delta s_{p-1} \\ \Delta\lambda \end{bmatrix} = - \begin{bmatrix} g_o \\ g_1 \\ \vdots \\ \\ g_p \end{bmatrix}$$

where $\Delta \underset{\sim}{s}_i$ is the correction to be made to $\underset{\sim}{s}_i$ and $\Delta \underset{\sim}{\lambda}$ is the correction to be made to $\underset{\sim}{\lambda}$.

It is easy to generalize multiple shooting to treat general nonlinear conditions of the form $c_i(t_i, \underset{\sim}{y}(t_i), \underset{\sim}{\lambda}) \cong 0$, $i=1,2,\ldots$ ($\cong$ means solve in a least squares sense).

3. Reducing the Number of Sets of Variational Equations

Let T_i be a nonsingular matrix such that

$$(3.1)\qquad T_i \Delta \underset{\sim}{s}_i = \begin{bmatrix} T_{1,1}^{(i)} & T_{1,2}^{(i)} \\ T_{2,1}^{(i)} & T_{2,2}^{(i)} \end{bmatrix} \begin{bmatrix} \Delta \underset{\sim}{s}_i^{(1)} \\ \Delta \underset{\sim}{s}_i^{(2)} \end{bmatrix} = \begin{bmatrix} \Delta \underset{\sim}{\sigma}_i^{(1)} \\ \Delta \underset{\sim}{\sigma}_i^{(2)} \end{bmatrix} = \Delta \underset{\sim}{\sigma}_i,$$

$$(3.2)\qquad T_i^{-1} = \begin{bmatrix} \overline{T}_{1,1}^{(i)} & \overline{T}_{1,2}^{(i)} \\ \overline{T}_{2,1}^{(i)} & \overline{T}_{2,2}^{(i)} \end{bmatrix},$$

where $\Delta \underset{\sim}{s}_i^{(1)}, \Delta \underset{\sim}{\sigma}_i^{(1)} \in \mathbb{R}^k$ and $\Delta \underset{\sim}{s}_i^{(2)}, \Delta \underset{\sim}{\sigma}_i^{(2)} \in \mathbb{R}^{n-k}$. Our goal is to find transformation matrices T_i such that $\Delta \underset{\sim}{\sigma}_i^{(2)}$ is immediately specified and only $\Delta \underset{\sim}{\sigma}_i^{(1)}$ need be determined. In the process the number of unknowns to be solved for in each shot is reduced by n-k with a corresponding reduction in the number of sets of variational equations which need to be integrated.

The case i=0 is special, but still serves as a useful introduction for what follows. Substituting into the first equation in (2.3), with $\sigma_o = T_o s_o$,

$$(3.3)\qquad A\underset{\sim}{s}_o - \underset{\sim}{y}_a = AT_o^{-1} T_o \underset{\sim}{s}_o - \underset{\sim}{y}_a = AT_o^{-1} \underset{\sim}{\sigma}_o - \underset{\sim}{y}_a = \underset{\sim}{0}.$$

Let $A=[A_1 A_2]$, and select $\overline{T}_{1,1}^{(0)}$ and $\overline{T}_{2,1}^{(0)}$ so that $A_1 \overline{T}_{1,1}^{(0)} + A_2 \overline{T}_{2,1}^{(0)} = 0$. Then eq. (3.3) can be written

$$(3.4) \qquad [A_1A_2]\begin{bmatrix}\bar{T}^{(0)}_{1,1} & \bar{T}^{(0)}_{1,2}\\ \bar{T}^{(0)}_{2,1} & \bar{T}^{(0)}_{2,2}\end{bmatrix}\begin{bmatrix}\underset{\sim}{\sigma}_o^{(1)}\\ \underset{\sim}{\sigma}_o^{(2)}\end{bmatrix} = [OL]\begin{bmatrix}\underset{\sim}{\sigma}_o^{(1)}\\ \underset{\sim}{\sigma}_o^{(2)}\end{bmatrix} = \underset{\sim}{\gamma}_a$$

where L is an $(n-k)\times(n-k)$ nonsingular matrix. (Computationally it is easy to determine the $\bar{T}^{(0)}_{i,j}$ so that L is lower triangular. With most software it will be most convenient to transform A to [LO] instead of [OL], thus interchanging $\underset{\sim}{\sigma}_o^{(1)}$ and $\underset{\sim}{\sigma}_o^{(2)}$.) Clearly $\underset{\sim}{\sigma}_o^{(2)}$ is the solution of

$$(3.5) \qquad L\underset{\sim}{\sigma}_o^{(2)} = \underset{\sim}{\gamma}_a$$

and $\underset{\sim}{\sigma}_o^{(1)}$ can take any desired values without violating eq. (3.4). With $\underset{\sim}{\sigma}_o^{(2)}$ determined by eq. (3.5) we need solve only for $\underset{\sim}{\sigma}_o^{(1)}$ (in place of $\underset{\sim}{s}_o$) in eq. (2.3). Examining the first two sets of rows of eq. (2.6) we find

$$(3.6) \qquad \begin{bmatrix}AT_o^{-1} & 0 & \cdots & 0\\ Y_o(x_1,\underset{\sim}{\lambda})T_o^{-1} & -I & \cdots & Y_o^{(\lambda)}(x_1,\underset{\sim}{\lambda})\end{bmatrix}\begin{bmatrix}T_o\Delta\underset{\sim}{s}_o\\ \Delta\underset{\sim}{s}_1\\ \vdots\\ \Delta\underset{\sim}{\lambda}\end{bmatrix} = -\begin{bmatrix}\underset{\sim}{g}_o\\ \underset{\sim}{g}_1\end{bmatrix}$$

which can be rewritten as

$$(3.7) \quad \begin{bmatrix}0 & L & 0 \cdots 0\\ Y_o(x_1,\underset{\sim}{\lambda})\begin{bmatrix}\bar{T}^{(0)}_{1,1}\\ \bar{T}^{(0)}_{2,1}\end{bmatrix} & Y_o(x_1,\underset{\sim}{\lambda})\begin{bmatrix}\bar{T}^{(0)}_{1,2}\\ \bar{T}^{(0)}_{2,2}\end{bmatrix} & -I \cdots Y_o^{(\lambda)}(x_1,\underset{\sim}{\lambda})\end{bmatrix}\begin{bmatrix}\Delta\underset{\sim}{\sigma}_o^{(1)}\\ \Delta\underset{\sim}{\sigma}_o^{(2)}\\ \Delta\underset{\sim}{s}_1\\ \vdots\\ \Delta\underset{\sim}{\lambda}\end{bmatrix} = -\begin{bmatrix}\underset{\sim}{g}_o\\ \underset{\sim}{g}_1\end{bmatrix}$$

With $\underset{\sim}{\sigma}_o^{(2)}$ satisfying eq. (3.5), $\underset{\sim}{g}_o=\underset{\sim}{0}$, and thus from eq. (3.7) $\Delta\underset{\sim}{\sigma}_o^{(2)}=0$. Thus the first set of rows and second set of columns of the matrix in (3.7) can be eliminated. Let

$$(3.8) \qquad V_i(t,\underset{\sim}{\lambda}) = Y_i(t,\underset{\sim}{\lambda}) \begin{bmatrix} \bar{T}^{(i)}_{1,1} \\ \bar{T}^{(i)}_{2,1} \end{bmatrix}$$

Thus V_i has k columns, whereas Y_i has n columns. From eq. (2.4)

$$(3.9) \qquad \frac{dV_i}{dt} = \frac{\partial \underset{\sim}{f}}{\partial \underset{\sim}{y}} V_i, \quad V_i(x_i,\underset{\sim}{\lambda}) = \begin{bmatrix} \bar{T}^{(i)}_{1,1} \\ \bar{T}^{(i)}_{2,1} \end{bmatrix}, \text{ and } V_i = \frac{\partial \underset{\sim}{g}_i}{\partial \underset{\sim}{\sigma}_i^{(1)}} .$$

Making the changes indicated above to (3.7) and appending the third set of rows from eq. (2.6) we find

$$(3.10) \qquad \begin{bmatrix} V_o(x_1,\underset{\sim}{\lambda}) & -I & & \cdots & Y_o^{(\lambda)}(x_1,\underset{\sim}{\lambda}) \\ & Y_1(x_2,\lambda) & -I & \cdots & Y_1^{(\lambda)}(x_2,\underset{\sim}{\lambda}) \end{bmatrix} \begin{bmatrix} \Delta\underset{\sim}{\sigma}_o^{(1)} \\ \Delta\underset{\sim}{s}_1 \\ \cdot \\ \cdot \\ \cdot \\ \Delta\underset{\sim}{\lambda} \end{bmatrix} = - \begin{bmatrix} \underset{\sim}{g}_1 \\ \underset{\sim}{g}_2 \end{bmatrix}$$

Let T_1 be a matrix such that

$$(3.11) \qquad T_1 V_o(x_1,\underset{\sim}{\lambda}) = \begin{bmatrix} U_1 \\ O \end{bmatrix}$$

where U_1 is a k×k matrix. Multiplying the first set of rows by T_1, replacing $\Delta\underset{\sim}{s}_1$ with $\Delta\underset{\sim}{\sigma}_1 = T_1\Delta\underset{\sim}{s}_1$, multiplying the second set of columns on the right by T_1^{-1} and breaking it further into two sets of columns, eq. (3.10) can be replaced by

$$(3.12) \qquad \begin{bmatrix} U_1 & -I & 0 & & & \cdots & T_1Y_o^{(\lambda)}(x_1,\underset{\sim}{\lambda}) \\ 0 & 0 & -I & & & & \\ 0 & V_1 & Y_1(x_2,\underset{\sim}{\lambda}) \begin{bmatrix} \bar{T}^{(1)}_{1,2} \\ \bar{T}^{(1)}_{2,2} \end{bmatrix} & -I & & \cdots & Y_1^{(\lambda)}(x_2,\underset{\sim}{\lambda}) \end{bmatrix} \begin{bmatrix} \Delta\underset{\sim}{\sigma}_o^{(1)} \\ \Delta\underset{\sim}{\sigma}_1^{(1)} \\ \Delta\underset{\sim}{\sigma}_1^{(2)} \\ \vdots \\ \Delta\underset{\sim}{\lambda} \end{bmatrix} = - \begin{bmatrix} T_1 g_1 \\ g_2 \end{bmatrix}$$

Let

$$(3.13)\qquad T_1 g_1 = \begin{bmatrix} \gamma_1^{(1)} \\ \gamma_1^{(2)} \end{bmatrix} = \gamma_1$$

The second set of rows of eq. (3.12) gives

$$(3.14)\qquad \begin{aligned} \Delta\sigma_1^{(2)} &= \gamma_1^{(2)} + \text{bottom of } T_1 Y_o^{(\lambda)}(x_1,\lambda)\Delta\lambda \\ &= \gamma_1^{(2)} + [T_{2,1}^{(1)}\ T_{2,2}^{(1)}]\ Y_o^{(\lambda)}(x_1,\lambda)\Delta\lambda \end{aligned}$$

We have implemented two different approaches from this point. Preliminary results indicate little difference in performance in the two approaches. In the same spirit as Gauss-Seidel iteration, the first approach uses "next iteration values" for some variables during the current iteration. It is a little easier to describe and requires less work. The second approach is mathematically equivalent to standard multiple shooting in the sense that it will give the same results if no errors are introduced by the initial value method or the linear equation solver but it does require the integration of an additional initial value system. The first approach begins with the observation that at this point in the algorithm we can satisfy eq. (3.14) using the current λ (i.e. $\Delta\lambda=0$) with

$$(3.15)\qquad \Delta\sigma_1^{(2)} = \gamma_1^{(2)}.$$

(Any other choice to solve eq. (3.14) involves $\Delta\lambda\neq 0$ which wouldn't help if λ were not present and would create additional complications at later breakpoints.) Having made this choice there is nothing to force eq. (3.14) to have a small residual when this point is reached in the next iteration. But we get this effect on the next iteration by once again selecting $\Delta\sigma_1^{(2)}$ using eq. (3.15).

The term

$$(3.16)\qquad Y_1(x_2,\lambda)\begin{bmatrix} \bar{T}_{1,2}^{(1)} \\ \bar{T}_{2,2}^{(1)} \end{bmatrix}\Delta\sigma_1^{(2)}$$

from the third set of equations in (3.12) is meant to

approximate

$$(3.17)\qquad g_2\left(s_1+T_1\begin{pmatrix}0\\ \Delta g_1^{(2)}\end{pmatrix},s_2\right)-g_2(s_1,s_2) = \hat{g}_2-g_2.$$

Thus instead of using this approximation (3.16), we replace it with the true value $\hat{g}_2-g_2$ with the result that (3.16) is deleted in eq. (3.12) and g_2 is replaced by $\hat{g}_2$. Thus instead of computing g_i for $i=2,3,\ldots,p$, $\hat{g}_i$ is computed by using

$$s_i+T_i^{-1}\begin{pmatrix}0\\ \Delta g_i^{(2)}\end{pmatrix}$$

for the initial value of $y_i(x_i,\lambda)$ instead of using s_i.

After dropping the 2nd set of rows and 3rd set of columns of eq. (3.12), and appending the next set of rows, we obtain:

$$(3.18)\qquad \begin{bmatrix} U_1 & -I & 0 & \cdots & [T_{1,1}^{(1)}\ T_{1,2}^{(2)}]Y_o^{(\lambda)}(x_1,\lambda) \\ 0 & V_1(x_2,\lambda) & -I & \cdots & Y_1^{(\lambda)}(x_2,\lambda) \\ 0 & 0 & Y_2(x_3,\lambda) & \cdots & Y_2^{(\lambda)}(x_3,\lambda) \end{bmatrix} \begin{bmatrix} \Delta g_o^{(1)} \\ \Delta g_1^{(1)} \\ \Delta s_2 \\ \vdots \\ \Delta\lambda \end{bmatrix} = -\begin{bmatrix} \gamma_1^{(1)} \\ \hat{g}_2 \\ g_3 \end{bmatrix}$$

Proceeding as we did following eq. (3.10), choose T_2 such that

$$T_2V_1(x_2,\lambda) = \begin{bmatrix} U_2 \\ 0 \end{bmatrix};$$

multiply the second set of rows by T_2; the third set of columns is multiplied from the right by T_2^{-1}; γ_2 is defined by

$$\gamma_2 = T_2\hat{g}_2 = \begin{bmatrix} \gamma_2^{(1)} \\ \gamma_2^{(2)} \end{bmatrix};$$ $\Delta g_2^{(2)}$ is set to $\gamma_2^{(2)}$; the bottom part of the 2nd set of rows and the right part of the 3rd set of columns are dropped; append the next set of rows; etc.

Thus eq. (2.6) is replaced by eq. (3.5) and

$$(3.19)\quad \begin{bmatrix} U_1 & -I & & & [T_{1,1}^{(1)} T_{1,2}^{(1)}]Y_o^{(\lambda)} \\ & \ddots & \ddots & & \vdots \\ & & U_{p-1} & -I & [T_{1,1}^{(p-1)} T_{1,2}^{(p-1)}]Y_{p-2}^{(\lambda)} \\ & & & BV_{p-1}(x_p,\lambda) & BY_{p-1}^{(\lambda)} \end{bmatrix} \begin{bmatrix} \Delta s_o^{(1)} \\ \vdots \\ \Delta s_{p-1}^{(1)} \\ \Delta\lambda \end{bmatrix} = - \begin{bmatrix} \gamma_1^{(1)} \\ \vdots \\ \gamma_{p-1}^{(1)} \\ B\hat{y}_{p-1}(x_p,\lambda) \\ -\gamma_b \end{bmatrix}$$

For the second approach we introduce some additional notation. Let

$$(3.20)\qquad \rho_i(t,\lambda) = Y_{i-1}(t,\lambda)\begin{bmatrix} \bar{T}_{1,2}^{(i-1)} \\ \bar{T}_{2,2}^{(i-1)} \end{bmatrix} \gamma_{i-1}^{(2)}$$

which is computed from the system of n differential equations

$$(3.21)\quad \frac{d}{dt}\rho_i(t,\lambda) = \frac{\partial f}{\partial y}\rho_i(t,\lambda),\ \rho_i(x_{i-1},\lambda) = \begin{bmatrix} \bar{T}_{1,2}^{(i-1)} \\ \bar{T}_{2,2}^{(i-1)} \end{bmatrix} \gamma_{i-1}^{(2)},\ t\in[x_{i-1},x_i],$$

$i=2,3,\ldots,p-1$. The solution of the equations (3.21) is the additional work required by the second approach. Let γ_1 be defined as in eq. (3.13),

$$(3.22)\qquad \gamma_i = T_i[s_i + \rho_i(x_i,\lambda)],\quad i=2,3,\ldots,p-1,$$

$$(3.23)\qquad \gamma_p = B[y_{p-1}(x_p,\lambda) + \rho_p(x_p,\lambda)] - \gamma_b.$$

Finally, let $W_o(t,\lambda) = Y_o^{(\lambda)}(t,\lambda)$ and

$$(3.24)\ W_i(t,\lambda) = Y_i(t,\lambda)\begin{bmatrix} \bar{T}_{1,2}^{(i)} \\ \bar{T}_{2,2}^{(i)} \end{bmatrix}[T_{2,1}^{(i)} T_{2,2}^{(i)}]W_{i-1}(x_i,\lambda) + Y_i^{(\lambda)}(t,\lambda), i=1,2,\ldots p-1.$$

The W_i can be computed with essentially the same cost as

the $Y_i^{(\lambda)}$ using

$$\text{(3.25)} \quad \frac{d}{dt}W_i(t,\underset{\sim}{\lambda}) = \frac{\partial \underset{\sim}{f}}{\partial \underset{\sim}{y}} W_i(t,\lambda) + \frac{\partial \underset{\sim}{f}}{\partial \underset{\sim}{\lambda}}, \quad t \varepsilon [x_i, x_{i+1}], \text{ with}$$

$$W_i(x_i,\underset{\sim}{\lambda}] = \begin{bmatrix} \bar{T}^{(i)}_{1,2} \\ \bar{T}^{(i)}_{2,2} \end{bmatrix} [T^{(i)}_{2,1} T^{(i)}_{2,2}] W_{i-1}(x_i,\underset{\sim}{\lambda}).$$

Substituting the value of $\Delta \underset{\sim}{g}_1^{(2)}$ defined by eq. (3.14) into the third row of eq. (3.12) and using (3.22), (3.20), and (3.24), we obtain

$$\text{(3.26)} \qquad V_1 \Delta \underset{\sim}{g}_1^{(1)} - \Delta \underset{\sim}{s}_2 + W_1(x_2,\underset{\sim}{\lambda}) \Delta \underset{\sim}{\lambda} = -T_2^{-1} \gamma_2 .$$

Note that, from (3.22), the right side of eq. (3.26) is the same for all nonsingular T_2.

Using (3.26) and (3.14) we no longer need the second set of rows and third set of columns of eq. (3.12). Taking advantage of this and appending the next set of rows, we have

$$\text{(3.27)} \begin{bmatrix} U_1 & -I & 0 & \cdots & [T^{(1)}_{1,1} T^{(1)}_{1,2}] W_o(x_1,\underset{\sim}{\lambda}) \\ 0 & V_1(x_2,\underset{\sim}{\lambda}) & -I & \cdots & W_1(x_2,\lambda) \\ 0 & 0 & Y_2(x_3,\underset{\sim}{\lambda}) & \cdots & Y_2^{(\lambda)}(x_3,\underset{\sim}{\lambda}) \end{bmatrix} \begin{bmatrix} \Delta \underset{\sim}{g}_o^{(1)} \\ \Delta \underset{\sim}{g}_1^{(1)} \\ \Delta \underset{\sim}{s}_2 \\ \vdots \\ \Delta \underset{\sim}{\lambda} \end{bmatrix} = - \begin{bmatrix} \underset{\sim}{\gamma}_1^{(1)} \\ T_2^{-1} \underset{\sim}{\gamma}_2 \\ \underset{\sim}{g}_3 \end{bmatrix}$$

Continuing in this way as we did before, choose T_2 such that $T_2 V_1(x_2,\underset{\sim}{\lambda}) = \begin{bmatrix} U_2 \\ 0 \end{bmatrix}$; multiply the second set of rows by T_2; multiply the third set of columns on the right by T_2^{-1}; define γ_2 and W_2 by equations (3.22) and (3.24); let $\Delta\sigma_2^2$ be defined by

$$\Delta\sigma_2^{(2)} = \underset{\sim}{\gamma}_2^{(2)} + [T^{(2)}_{2,1} T^{(2)}_{2,2}] W_1(x_2,\underset{\sim}{\lambda}) \Delta\lambda$$

the bottom part of the 2nd set of rows and the right part of the 3rd set of columns are dropped; the next set of rows appended; etc. Thus instead of eq. (3.19) we obtain

$$\text{(3.28)}\quad \begin{bmatrix} U_1 & -I & & & [T_{1,1}^{(1)}T_{1,2}^{(1)}]W_o(x_1,\lambda) \\ & \ddots & & & \\ & & U_{p-1} & -I & [T_{1,1}^{(p-1)}T_{1,2}^{(p-1)}]W_{p-1}(x_{p-1},\lambda) \\ & & & BV_{p-1}(x_p,\lambda) & BW_{p-1}(x_p,\lambda) \end{bmatrix} \begin{bmatrix} \Delta\sigma_o^{(1)} \\ \vdots \\ \Delta\sigma_{p-1}^{(1)} \\ \Delta\lambda \end{bmatrix} = - \begin{bmatrix} \gamma_1^{(1)} \\ \vdots \\ \gamma_{p-1}^{(1)} \\ \gamma_p \end{bmatrix}$$

4. The Interface to the Nonlinear Equation Solver

Both equations (3.19) and (3.28) have the unknowns $\Delta\lambda$, $\Delta\sigma_i^{(1)}$, $i=0,1,\ldots,p-1$. Once again there seem to be two approaches. The first approach, and the only one we have implemented, informs the nonlinear equation solver only of the variables λ and $\sigma_i^{(1)}$ where $\Delta\sigma_i = T_i\Delta s_i$. In order for this to be correct these variables must be the same from iteration to iteration (i.e. $T_{11}^{(i)}$ and $T_{12}^{(i)}$ must not vary from one iteration to the next). In order to accomplish this we have chosen T_i to be of the form:

$$\text{(4.1)}\qquad T_i = \begin{bmatrix} I & 0 \\ M_i & I \end{bmatrix} P_i$$

where P_i is a permutation matrix selected on the first iteration to make M_i "small" (and thus give T_i a low condition number), and M_i is allowed to change from one iteration to the next. It is easy to verify that

$$\text{(4.2)}\qquad T_i^{-1} = P_i^T \begin{bmatrix} I & 0 \\ -M_i & I \end{bmatrix}$$

Although this is usually satisfactory, our code will sometimes do a restart with a new P_i and new breakpoints.

In our early work, we overlooked the possibility of transforming back to the original coordinates, i.e. s_i and λ, before reference is made to the nonlinear equation solver. Although it requires more work we believe this approach to be the more promising since it allows the T_i to be orthogonal matrices that change from iteration to iteration, which in turn will lead to nonlinear problems with better conditioned iteration matrices. It also simplifies treating user constraints on the solution if the underlying nonlinear equation solver has such a capability, as ours does.

The form of eq. (3.19) in the case p=3 when this approach is used is given by

$$
(4.3)\quad
\begin{bmatrix}
A & 0 & 0 & 0 \\
\begin{bmatrix} U_1 & 0 \\ 0 & -\alpha I \end{bmatrix} T_o & \begin{bmatrix} -I & 0 \\ 0 & 0 \end{bmatrix} T_1 & \begin{bmatrix} 0 \end{bmatrix} & \begin{bmatrix} T^{(1)}_{1,1} & T^{(1)}_{1,2} \\ 0 & 0 \end{bmatrix} Y_o^{(\lambda)}(x_1,\underset{\sim}{\lambda}) \\
\begin{bmatrix} 0 \end{bmatrix} & \begin{bmatrix} -U_2 & 0 \\ 0 & -\alpha I \end{bmatrix} T_1 & \begin{bmatrix} -I & 0 \\ 0 & 0 \end{bmatrix} T_2 & \begin{bmatrix} T^{(2)}_{1,1} & T^{(2)}_{1,2} \\ 0 & 0 \end{bmatrix} Y_1^{(\lambda)}(x_2,\underset{\sim}{\lambda}) \\
\begin{bmatrix} 0 \end{bmatrix} & \begin{bmatrix} 0 \end{bmatrix} & BV_2(x_3,\underset{\sim}{\lambda})T_2 & BY_2^{(\lambda)}(x_3,\underset{\sim}{\lambda})
\end{bmatrix}
\begin{bmatrix} \Delta\underset{\sim}{s}_o \\ \Delta\underset{\sim}{s}_1 \\ \Delta\underset{\sim}{s}_2 \\ \Delta\underset{\sim}{\lambda} \end{bmatrix}
= -
\begin{bmatrix} A\underset{\sim}{s}_o-\underset{\sim}{\gamma}_a \\ \underset{\sim}{\gamma}_1^{(1)} \\ \alpha\underset{\sim}{\gamma}_1^{(2)} \\ \underset{\sim}{\gamma}_2 \\ \alpha\underset{\sim}{\gamma}_2 \\ B\hat{y}_2(x_3,\underset{\sim}{\lambda}) \\ -\underset{\sim}{\gamma}_b \end{bmatrix}
$$

where $\hat{\underset{\sim}{y}}_2$ satisfies $\hat{\underset{\sim}{y}}_2'=f(t,\hat{y}_2,\lambda)$, $\hat{\underset{\sim}{y}}_2(x_2,\underset{\sim}{\lambda})=s_2+T_2\begin{bmatrix} 0 \\ \Delta\underset{\sim}{\sigma}_2^{(2)} \end{bmatrix}$, and the parameter α is selected enough bigger than 1 so that the last n-k components of $T_i\Delta\underset{\sim}{s}_i$ are close to $\Delta\underset{\sim}{\sigma}_i^{(2)}=\underset{\sim}{\gamma}_i^{(2)}$, but not so big that conditioning of the matrix is unduly effected. (We do not expect to solve the equations exactly since the norm of correction is restricted by the nonlinear solver.) Extending this approach to other values of p or to equations (3.28), (3.26) is done in the obvious way.

5. Getting Started on Nonlinear Problems

When starting with a poor initial guess with a Newton (or for least squares a Gauss-Newton) iteration requires that the size of corrections be restricted in some way. Although such a facility is part of our nonlinear solver we circumvent this feature by getting a direct estimate for $\Delta\underset{\sim}{\sigma}_i^{(2)}$. This caused failure on many problems.

An automatic continuation method seems to be an attractive way to solve this problem and this is recommended as a substitute (or in addition) to what we describe below. We have coped with bad initial guesses using a method we call progressive multiple shooting. By progressive multiple shooting we mean any method that establishes initial values for the multiple shooting in progression, by doing integrations over only a part of the interval. We recommend the following variant of this method.

1. At the conclusion of the i-th shot (determined by a user or program selected x_i) an initial guess of the starting value $\underset{\sim}{s}_i$ for the next shot is supplied.
2. Replace this $\underset{\sim}{s}_i$ with $\hat{\underset{\sim}{s}}_i=[I-V_{i-1}(x_i,\underset{\sim}{\lambda})V^+_{i-1}(x_i,\underset{\sim}{\lambda})]s_i$, where V^+_{i-1} is the pseudo inverse of V_{i-1}. Thus $\hat{\underset{\sim}{s}}_i$ is the projection of s_i into the space orthogonal to V_{i-1}.
3. Iterate to get a value of $\underset{\sim}{\sigma}^{(2)}_{i-1}$ that gives a $\underset{\sim}{g}_i$ within $\sim$50% of its minimum value. (Perhaps changing $\underset{\sim}{s}_i$ to $\hat{\underset{\sim}{s}}_i$ after each iteration.)
4. Continue as described earlier with the (i+1)-st shot, using $\hat{\underset{\sim}{s}}_i$ in place of $\underset{\sim}{s}_i$.

In the code step 2 is not used, thus $\hat{\underset{\sim}{s}}_i \equiv \underset{\sim}{s}_i$. We believe the small additional amount of work to compute $\hat{\underset{\sim}{s}}_i$ will pay dividends. Also, in our code in step 3, $\underset{\sim}{\chi}^{(2)}_j$ is solved for, for j=0,1,...,i-1. With the addition of step 2, this additional work is probably not needed.

References

[1] David D. Morrison, James D. Riley, and John F. Zancanaro, "Multiple shooting method for two-point boundary value problems", Comm. ACM 5 (1962), pp. 613-614.

[2] Herbert B. Keller, Numerical Methods for Two-Point Boundary Value Problems, Ginn-Blaisdell, Waltham, Mass. 1968.

[3] Herbert B. Keller, Numerical Solution of Two-Point Boundary Value Problems, Regional Conference Series in Applied Mathematics, No. 24, Society for Industrial and Applied Mathematics, Philadelphia, Pa. 1976.

[4] Michael R. Osborne, "On the Numerical Solution of Boundary Value Problems for Ordinary Differential Equations", Information Processing '74, North Holland, Amsterdam, 1974, pp. 673-677.

[5] Peter Deuflhard, "Recent advances in multiple shooting techniques", in Computational techniques for ordinary differential equations, Academic Press, New York, 1980, pp. 217-272.

[6] John H. George and Robert W. Gunderson, "Conditioning of linear boundary value problems", BIT 12, (1972), pp. 172-181.

[7] Marianela Lentini, Michael R. Osborne and Robert D. Russell, "The close relationships between methods for solving two-point boundary value problems", to appear in SIAM J. Numer. Anal.

[8] S.D. Conte, "The numerical solution of boundary value problems", SIAM Rev. 8 (1966), pp. 309-321.

[9] M.R. Scott and H.A. Watts, "Computational solution of linear two point boundary value problems via orthonormalization", SIAM J. Numer. Anal. 14, (1977), pp. 40-70.

[10] Michael R. Osborne, "The stabilized march is stable", SIAM J. Numer. Anal. 16, (1979), pp. 923-933.

[11] R.M.M. Mattheij and G.W.M. Staarink, "An efficient algorithm for solving general linear two point BVP", Report 8220, Math. Inst. Catholic University, Nijmegen, (1982).

Progress in Scientific Computing, Vol. 5
Numerical Boundary Value ODEs

THE SPLINE-COLLOCATION AND THE SPLINE-GALERKIN METHODS FOR ORR-SOMMERFELD PROBLEM

A.G.Sleptsov

1. Recently the projectional mesh methods of solving the boundary value problems for ordinary differential equations are intensively developing. The theoretical and practical aspects of the spline-collocation method have been examined by R.D.Russell and L.F.Shampine [15] . C. de Boor and B.Swartz [5] have studied a question of choosing the points in the spline-collocation method. They have shown that the greatest rate will be achieved by the choice of the Gaussian points as the collocation points. B.P.Kolobov and A.G.Sleptsov [11,12] have made use of the following method for solving the problem

$$Ly = f, \ (B_0 y)(a) = 0, \ (B_1 y)(b) = 0 \qquad (1)$$

where L is linear mth order differential operator. Let $x_1=a < x_2 < \dots < x_{N+1}=b$, $n > 0$ is some interger. For each $k=1, \dots, N-m$ the approximate solution will be a polynomial of the degree $m+n-1$ on the segment $[x_k, x_{k+m}]$. m coefficients of this polynomial are determined by the conditions $u_k(x_i)=v_i$, $i=k, \dots, k+m-1$, the rest of the coefficients are determined by the collocation equations. The equalities $v_{k+m}=u_k(x_{k+m})$, $k=1, \dots, N-m$, give N-m linear equations to determine the N unknowns v_i. The boundary conditions yield m more equations. It is clear that $u_k(x)$ and $u_{k+1}(x)$ are the approximate solutions of the proble blem (1) on the intersecting intervals $[x_k, x_{k+1}]$ and $[x_{k+1}, x_{k+m+1}]$ and they coincide at the m points: $x_{k+1}, \dots$

x_{k+m}, therefore this method may be called the method with multi-point matching.

The substantiation of this method has been given by A.G.Sleptsov [18-20] . Recently the collocation method with polynomials on the intersecting intervals was considered by E.J.Doedel [9] , R.E.Lynch and J.R.Rice [13] . The universality and high effectiveness of the spline-collocation method was shown in papers [1,2,6,7,16,17] . Therefore it was natural to employ this method for developing software for solving the problems of hydrodynamic stability.

In the present paper the algorithms of solving the Orr-Sommerfeld problem which are based on the spline-collocation method and the spline-Galerkin method are described. The implementation of these methods is based on singling out the local solutions as it was done earlier in papers [11,12,18-20] ; that has proved to be considerably more effective as compared to the others approaches, as was shown in paper [3]. The results of computations indicate that the representation of the solution as piecewise polynomials on non-intersecting intervals is more effective than those considered before [11,12,18,20] . The algorithm performs 5-10 times faster. The spline-colocation method gives comparable accuracy to the spline-Galerkin method, but its implementation is simpler.

2. The linear problem of the hydrodynamic stability of the plane parallel flows of the fluid is formulated (see e.g. [4,8]) as the eigenvalue problem for the Orr-Sommerfeld equation that is defined on the segment $[0,1]$ with some uniform conditions

$$Ly = \lambda My \qquad (2)$$

$$\sum_{j=1}^{4} \gamma_{ij} y^{(j-1)}(0) = 0, \; i=1,2, \quad \sum_{j=1}^{4} \gamma_{ij} y^{(j-1)}(1) = 0, \; i=3,4, \qquad (3)$$

where $Ly \equiv y^{(4)} + \sum_{i=1}^{4} p_i y^{(i-1)}$, $My \equiv y'' - \alpha^2 y$, $p_1 = \alpha^4 + i\alpha^3 RU(x) + i\alpha RU''(x)$, $p_2 = \alpha^2 VR$, $p_3 = -2\alpha^2 - i\alpha RU(x)$,

$p_4=-VR$, α, R,V are the known parameters, U=U(x) is the known function, the coefficients γ_{ij} of the boundary conditions depend on the parameters α ,R and V and may depend on the parameter λ .

Let $x_1=0 < x_2 < \dots < x_N < x_{N+1}=1$ be some mesh on the segment $[0,1]$. The solution of the problem (2),(3) will be sought in the form of a piecewise-polynomial function of class $C^3[0,1]$, the degree of the polynomial v_k being equal to n+3 on each segment $[x_k, x_{k+1}]$. Let us dwell on the question of the definition of polynomials in detail. Map the segment $[x_k, x_{k+1}]$ by the transformation $x \longmapsto \xi =(x-x_k)/h_k-1$ onto the segment $[-1,1]$. Then Eq.(1) is transformed into the equation

$$\bar{L}\bar{y} = \lambda \bar{M}\bar{y}, \tag{4}$$

where $\bar{y}(\xi)=y(x)$. Let u_{ki}, i=1,...,4, be the polynomials of the degree n+3 that satisfy the boundary conditions

$$\begin{aligned}
&u_{k1}(-1)=1, \quad u'_{k1}(-1)=0, \quad u_{k2}(-1)=0, \quad u'_{k2}(-1)=h_k,\\
&u_{k3}(1)=1, \quad u'_{k3}(1)=0, \quad u_{k4}(1)=0, \quad u'_{k4}(1)=h_k, \qquad (5)\\
&u^{(i)}_{kj}(1)=0, \quad j=1,2, \quad u^{(i)}_{kj}(-1)=0, \quad j=3,4, \quad i=0,1.
\end{aligned}$$

The rest of the coefficients $u_{ki}(x)$ are defined from the collocation equations or from the Galerkin's equations written for Eq.(4). Denote $y(x_k)=y_{2k-1}$, $y'(x_k)=y_{2k}$, and write the polynomial $\bar{v}_k(\xi)$ in the form

$$\bar{v}_k(\xi)=y_{2k-1}u_{k1}(\xi)+y_{2k}u_{k2}(\xi)+y_{2k+1}\, u_{k3}(\xi)+y_{2k+2}u_{k4}(\xi) \tag{6}$$

It follows from the definition of polynomials $u_{ki}(\xi)$ that the function $v(x)=v_k(x)=\bar{v}_k(\xi)$ at $x\in[x_k, x_{k+1}]$ belongs to the class $C^1[0,1]$ at any value of the parameters y_i. Determine these parameters from the continuity condition of the second and third derivatives of the function $v(x)$ at the point x_k for any point x_k, $k=2,\dots, N$:

$$v^{(i)}(x_k-0)=v^{(i)}(x_k+0), \quad i=2,3 \tag{7}$$

Taking into account the equality (6) a system of equations (7) can be written as follows

$$\sum_{j=2k-3}^{2k+2} a_{ij}(\lambda)y_j=0, \quad i=2k-1,\ 2k,\ k=2,\dots, N \tag{8}$$

where

$$a_{1,2k-3}(\lambda)=-r_k^i\, u_{k-1,1}^{(i)}(1), \quad a_{1,2k-2}(\lambda)=-r_k^i\, u_{k-1,2}^{(i)}(1),$$

$$a_{1,2k-1}(\lambda)=u_{k1}^{(i)}(-1)-r_k^i\, u_{k-1,3}^{(i)}(1), \quad a_{1,2k}(\lambda)=u_{k2}^{(i)}(-1)-$$

$$-r_k^i\, u_{k-1}^{(i)}(1)$$

$$a_{1,2k+1}(\lambda)=u_{k3}^{(i)}(-1), \quad a_{1,2k+2}(\lambda)=u_{k4}^{(i)}(-1), \quad 1=2k+i-3,$$

$$i=2,3, \quad r_k=\frac{h_{k-1}}{h_k}$$

The system (8) consists of $2(N-1)$ equations and contains $2(N+1)$ unknowns y_i, $i=1,\dots, 2N+2$. The boundary conditions (3) yield 4 more equations

$$\sum_{j=1}^{4} a_{ij}(\lambda)y_j=0, \quad i=1,2$$

$$\sum_{j=2N-1}^{2N+1} a_{Ij}(\lambda)y_j=0, \quad i=2N+1,\ 2N+2 \tag{9}$$

<u>Remark.</u> For determining the coefficients of polynomials u_{ki} at a fixed k and $i=1,\dots,4$ we derive systems that differ only by the right-hand sides, because polynomials u_{ki}, $i=1,\dots,4$, are approximate solutions of the equation (2) on the segment $[x_k, x_{k+1}]$ that satisfy different boundary conditions (4) Therefore these systems are solved simultaneously.

The eigenvalues λ of a system of equations (8), (9) are determined as zeros of the determinant Δ of this system with the help of secant method. It is assumed that the initial approximation for the eigenvalue is set.

The determinant of a system of equations (8), (9) may be found by Gauss elimination when selecting a main element in the line and preserving the structure of the matrix. The matrix A of the system has the structure

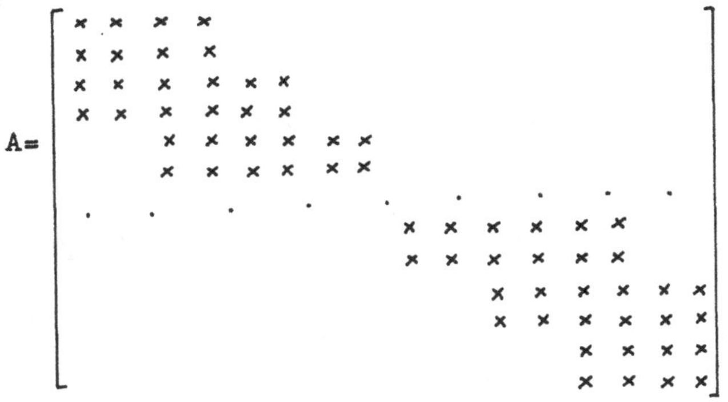

A=

Let us have a system of the form I before the 2k - 1 step

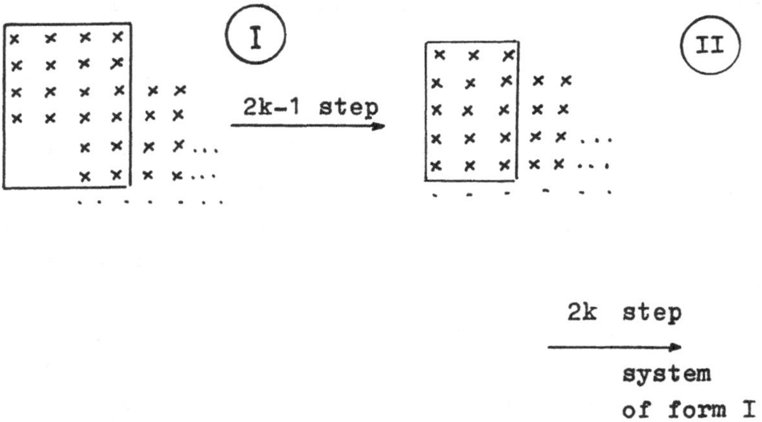

I

2k-1 step

II

2k step

system of form I

At the 2k - 1 and 2k steps all the transformations are performed only with the matrix elements contained in the frame. In the first line the element of the matrix with the largest module $|\bar{a}_{2k-1,i}| = \max |\bar{a}_{2k-1,j}|$ is chosen. If

$\bar{a}_{2k-1,i} = 0$, then the determinant Δ of the system of equations (8), (9) is equal to 0 and the calculation of the determinant is finished. If $\bar{a}_{2k-1,i} \neq 0$, then $\Delta := a_{2k-1,i} \cdot \Delta_{2k-2}$. We exclude the unknown with the number i and derive a system of the form II. In this system we perform similar transformations and derive again a system of the form I and the process is repeated for all k=1,..., 2N-2. Before the 2N - 1 step we have a 4 x 4 matrix whose determinant δ is computed in a standard way and finally obtain that the determinant Δ of the system (8), (9) is equal to

$$\Delta(\lambda) = \Delta_{2N-2}(\lambda) \cdot \delta(\lambda)$$

A similar algorithm for solving the systems with a band matrix was used by B.P.Kolobov [10] and V.N.Shepelenko.

In computations the mesh step $h_k = x_{k+1} - x_k$ was defined with the help of a cubic polynomial

$h_k = \sum_{i=1}^{4} b_i k^{i-1}$, the coefficients

b_i were determined empirically and when chosen for one model problem yielded good results for the other problems.

3. In order to verify the efficiency of various algorithms, we have solved the two well-known problems: the stability of the plane parallel Poiseuille flow [4,8, 14]and the stability of the Blasius profile in the boundary layer [4,8] . Table 1 shows the results of calculations of the eigenvalues for the Poiseuille flow, the GM-Galerkin method, TP, GP and MLP are the Tchebysov, Gaussian and Markov-Lobatto's points in the collocation method, respectively. All the figures with exception of the last one coincide with the corresponding figures of the exact solution. We can see from the Table 1 that the best results are obtained if the Galerkin method or the collocation method with the

TABLE 1. The comparison of the various methods for the calculation of eigenvalue for the Orr-Sommerfeld problem for Poiseuille flow, $\alpha = 0.65$, $R = 7.29 \cdot 10^5$ $= -i\alpha Rc$.

N	n	GM c_r	GM c_i	TP c_r	TP c_i	GP c_r	GP c_i	MLP c_r	MLP c_i
	3	0.12	0.002	0.12	-0.020	0.08	0.003	0.12	-0.02
	4	0.100	0.001	0.11	-0.020	0.09	-0.0006	0.12	-0.002
	5	0.091	0.002	0.0904	0.0022	0.0903	0.0012	0.095	-0.0006
8	6	0.0902	0.0015	0.0902	0.0011	0.09025	0.00138	0.0907	0.0017
	7	0.090244	0.001352	0.0903	0.0015	0.090244	0.001357	0.09022	0.0014
	8	0.090244	0.001352	0.09021	0.00134	0.09024412	0.00135328	0.090236	0.001349
	9	0.090244	0.001353	0.09025	0.00135	0.09024420	0.00135323	0.0902441	0.0013526
	10	0.090244	0.001353	0.090243	0.001354	0.09024418	0.00135326	0.09024420	0.00135327
	2	0.11	-0.001	0.11	-0.01	0.090	0.001	0.10	-0.03
	3	0.0906	0.0007	0.088	-0.0005	0.0902	0.0011	0.08	-0.001
	4	0.09028	0.00138	0.0906	0.0014	0.09023	0.00136	0.0904	0.0012
16	5	0.090244	0.001356	0.09020	0.00139	0.090244	0.001355	0.09028	0.00136
	6	0.090244	0.001353	0.090247	0.001345	0.09024423	0.00135329	0.0902436	0.0902436
	7			0.090244	0.001354	0.09024419	0.00135326	0.09024411	0.00135323
	8			0.090244	0.001353	0.09024418	0.00135326	0.09024418	0.00135326
	1	0.14	0.04	0.089	0.005	0.089	0.005	0.089	0.005
	2	0.0899	0.0010	0.089	0.002	0.0902	0.0012	0.088	0.007
	3	0.090249	0.001347	0.0901	0.0012	0.09023	0.00135	0.0900	0.0011
32	4	0.090244	0.001353	0.09025	0.00136	0.0902442	0.0013535	0.09026	0.00135
	5			0.090243	0.001354	0.09024419	0.00135326	0.0902443	0.0013532
	6			0.090244	0.001353	0.09024418	0.00135326	0.09024418	0.00135326

TABLE 2. Eigenvalues for the spline-Galerkin approximations of the Orr-Sommerfeld problem. Blasius flow. α=0.105, R=10^4

N	n	c_r	c_i	t
1	9	0.15	0.06	0.13
	10	0.16	0.04	0.14
2	7	0.23	0.02	0.19
	8	0.22	0.005	0.25
	9	0.226	0.003	0.26
	10	0.2274	0.0034	0.28
	10	0.22688	0.00392	0.28*
4	3	0.25	-0.004	0.11
	4	0.228	0.005	0.14
	5	0.2267	0.0041	0.16
	6	0.22689	0.00389	0.23
	7	0.22690	0.00395	0.25
	8	0.22687	0.00395	0.31
	8	0.226901	0.00393	0.30*
	9	0.22687	0.00394	0.38
	10	0.22687	0.00394	0.47
8	2	0.225	-0.004	0.16
	3	0.2271	0.0040	0.25
	4	0.22689	0.00387	0.28
	4	0.22684	0.00387	0.27*
	5	0.22691	0.00393	0.32
	5	0.22691	0.00392	0.30*
	6	0.226903	0.003917	0.43
	6	0.226903	0.003917	0.40*
16	1	0.221	0.005	0.19
	2	0.2269	0.0038	0.28
	3	0.22689	0.00391	0.38
	3	0.22687	0.00389	0.37*
	4	0.226904	0.003917	0.44
	5	0.226903	0.003917	0.56
32	1	0.2255	0.0038	0.14
	2	0.226901	0.00390	0.25
	3	0.226903	0.003917	0.33
	3	0.226903	0.003917	0.30*
64	1	0.2265	0.0038	0.28
	2	0.226903	0.003916	0.50

*) the Gaussian points; t is the time of the BESM-6 for one iteration in sec., $\lambda = -i\alpha Rc$

Gaussian knots are used. The collocation method with the matching of polynomials at one point exceeds in accuracy the collocation method with multipoint matching [11,12,18-20].

TABLE 3. Spline-collocation method with Gaussian points, the Poiseuille flow, $\alpha = 1$, $R=10^4$, $\lambda = -i\alpha Rc$.

N	n	c_r	c_i
16	4	0.23752649	0.00373967
8	6		
4	10		
16	7	0.96464251	-0.03518658
8	9		
16	7	0.93635178	-0.06325157
16	8	0.90805633	-0.09131286
16	9	0.87975570	-0.11937073
16	7	0.34910682	-0.12450198
16	9	0.8514494	-0.1474256
16	10	0.8231370	-0.1754781
16	6	0.1900592	-0.1828219
8	9		
16	10	0.794818	-0.203529
16	8	0.474901	-0.208734
32	7	0.76649	-0.23158(9)
16	9	0.36850	-0.23882
32	8	0.73812	-0.25969
16	10	0.58721	-0.26716
16	9	0.51292	-0.28662(3)
32	8	0.70887	-0.28765

The results of computations of the eigenvalue for the Blasius profile for one point of the (α,R)-plane are represented in Table 2.

Table 3 illustrates the eigenvalues which were obtained by S.A.Orszag [14], the number of intervals - N and the order of accuracy - n that allow one to get these eigenvalues by the spline-collocation method. In the Table the figure in the parentheses for two eigenvalues was taken from [14]. The results obtained in the present work were checked by computations performed on a mesh with the number of points doubled and a divergence from Orszag's results was obtained only in the last figure for two eigenvalues.

REFERENCES

[1] Ascher U., Christiansen J., Russell R.D. A collocation solver for mixed order systems of boundary value problems. "Math. Comp.", 1979, 33, 146, 659-679.

[2] Ascher U., Christiansen J., Russell R.D. Collocation software for boundary value ODEs. "ACM Transaction on mathematical Software", 1981, 7, 2, 209-222.

[3] Ascher U., Pruess S., Russell R.D. A spline basis selection for solving differential equations. SIAM J. Numer. Anal., 1983, v.20, N 7, 121-142.

[4] Betchov R., Criminale W.O., Stability of parallel flows, Academic Press, New York - London, 1967.

[5] De Boor C., Swartz B. Collocation at Gaussian points, SIAM J. Numer. Anal., 1978, 10, 3, 582-606.

[6] Christiansen J., Russell R.D. Deferred corrections using uncentered end formulas. "Numer. Math.", 1980, 35, 21-33.

[7] Christiansen J., Russell R.D. Error analysis for spline collocation methods with application to knot selection. "Math. Comp.", 1978, 32, 415-419.

[8] Гольдштик М.А., Штерн В.Н. Гидродинамическая устойчивость и турбулентность. Новосибирск, Наука, 1972.

[9] Doedel E.J. Finite difference collocation method for nonlinear two point boundary value problems. "SIAM J. Numer. Anal.", 1979, 16, 2, 173-185.

[10] Колобов Б.П. Метод немонотонной прогонки для решения систем линейных уравнений с m-диагональной матрицей ($m \geq 2$). - В сб.:"Численные методы механики сплошной среды", 1979, т. 10.

[11] Колобов Б.П., Слепцов А.Г. Новый метод построения разностных схем для обыкновенных дифференциальных уравнений и его применение к решению задачи. В сб.: "Численные методы механики сплошной среды", Новосибирск, 1972, 3, № 1, 61-77.

[12] Колобов Б.П., Слепцов А.Г. Устойчивость двухслойных течений жидкости.Изв.АН СССР, МЖГ, 1974, № 5, 155.

[13] Lynch R.E., Rice J.R. A high-order difference method for differential equations. "Math. Comp.", 1980, 34, 150, 333-372.

[14] Orszag S.A. Accurate solution of the Orr-Sommerfeld stability equation. "J. Fluid Mech.", 1971, 50, 4, 689-703.

[15] Russell R.D. A comparison of collocation and finite differences for two-point boundary value problems.

"SIAM J. Numer. Anal.", 1977, 14, 1.

[16] Russell R.D. Collocation for systems of boundary value problems. "Numer. Meth.", 1974, 23, 119-139.

[17] Russell R.D., Shampine L.F. A collocation method for boundary value problems, "Numer. Math.", 1972, 19, 1-28.

[18] Слепцов А.Г. Сходимость метода локальной коллокации для обыкновенных дифференциальных уравнений. "Ж. вычисл. матем. и матем. физ.", 1975, 15, № 6, 1447-1456.

[19] Слепцов А.Г. Сходимость метода локальной коллокации. - В сб.: "Численные методы механики сплошной среды", 1977, 8, № 7, 141-154.

[20] Слепцов А.Г. Метод локальной коллокации в проблеме собственных значений. - В сб.: "Численные методы механики сплошной среды", 1978, 9.

Progress in Scientific Computing, Vol. 5
Numerical Boundary Value ODEs

On the Simultaneous Use of Asymptotic and Numerical Methods to Solve Nonlinear Two Point Problems with Boundary and Interior Layers

Robert E. O'Malley, Jr.

The purpose of this paper is to provide a broad-brush survey concerning boundary value problems for certain systems of nonlinear singularly perturbed ordinary differential equations. The aim is to emphasize important and difficult open problems needing much more study, in terms of both mathematical and numerical analysis and computational experiments. The presentation will, regrettably, be removed from both direct applications and substantial achievements. Gradually, we will, however, become more specific and ultimately will discuss some currently tractible problems.

Let us first consider a general vector problem

$$\begin{aligned} \dot{z} &= f(z,t) \\ b\big(z(0),\ z(1)\big) &= 0\ , \end{aligned} \tag{1}$$

on a bounded interval $0 \le t \le 1$, whose solutions feature a finite number of narrow regions of rapid change at either endpoint or at interior points. (Thus, we eliminate consideration of systems involving persistent rapid oscillations.) The identification of different time scales of motion becomes most apparent when we can transform variables by setting

$$z = h(w,t) \quad \text{or} \quad w = j(z,t) \tag{2}$$

to obtain a new system of the form

$$\Omega(\varepsilon)\dot{w} = \ell(w,t,\varepsilon) \tag{3}$$

where, say, ℓ is everywhere bounded and

$$\Omega(\varepsilon) = \mathrm{diag}\left(I_{k_0}, \varepsilon_1 I_{k_1}, \ldots, \varepsilon_n I_{k_n}\right)$$

with n successively smaller positive parameters simultaneously tending to zero. Corresponding to (3), it is natural to split

$$w = \begin{pmatrix} w_0 \\ w_1 \\ \cdot \\ \cdot \\ \cdot \\ w_n \end{pmatrix}$$

into subvectors of slow and (potentially) progressively faster components. In practical applications, such aggregation (or decomposition) of variables can often be based on physical information. In electric power systems, for example, mechanical and electrical variables often evolve on vastly different time scales (cf. Chow (1982)). In fluid dynamics, on the other hand, one often uses dimensional analysis to provide nondimensional parameters which may take extreme values (cf. Van Dyke (1964)). An artform has developed concerning the appropriate scaling of dependent and independent variables, which has been systematized in a few situations. (Interesting recent studies include those by Mao (1982) in Stockholm and Nipp (1980) in Zurich, who used linear programming methods to determine the stretchings needed for certain initial value problems.) Kreiss has sought to numerically generate new variables, in terms of which solutions will depend smoothly (cf. Kreiss, Nichols, and Brown (1983)). Flaherty and

O'Malley (1984) have used Householder transformations to decouple slow and fast motion for certain quasilinear problems, while Mattheij and O'Malley (1984) have combined Riccati transformations, the power series of classical analytic theory, and multiple shooting to attack some linear problems. In using such transformations from a given problem to a more convenient canonical form, it is important that the original variables be easily recovered, since they are presumably of physical significance.

It is appropriate and necessary to review some classical theory (which is predominantly due to Tikhonov and Levinson (simultaneously) in the early 1950's). Let us first consider initial value problems for nonlinear systems

$$\begin{aligned} \dot{x} &= f(x,y,t,\varepsilon) \\ \varepsilon\dot{y} &= g(x,y,t,\varepsilon) \end{aligned} \qquad (4)$$

where x is an m vector of slow variables, y is an n vector of potentially fast variables, and ε is a small positive parameter tending toward zero. We'll presume plenty of smoothness and that the n eigenvalues of the Jacobian matrix g_y all have strictly negative real parts everywhere. Then (cf. Wasow (1965), Vasil'eva and Butuzov (1973), or O'Malley (1974)), there is an asymptotic solution of the form

$$\begin{aligned} x &= X(t,\varepsilon) + \varepsilon\xi(\tau,\varepsilon) \\ y &= Y(t,\varepsilon) + \eta(\tau,\varepsilon) \end{aligned} \qquad (5)$$

where $\binom{X}{Y}$ represents the outer solution and $\binom{\varepsilon\xi}{\eta}$ is an initial layer correction which decays to zero as the stretched fast variable

$$\tau = t/\varepsilon \qquad (6)$$

tends to infinity. In particular, note that η allows the components of y to converge nonuniformly from its outer limit $Y(t,0) = Y_0(t)$ to its prescribed initial value $y(0)$ as $\varepsilon \to 0$. Further, the limiting outer solution must satisfy the reduced system

$$\begin{aligned} \dot{X}_0 &= f(X_0,Y_0,t,0) \\ 0 &= g(X_0,Y_0,t,0) \end{aligned} \tag{7}$$

and the initial condition $X_0(0) = x(0)$. Since g_y is nonsingular, we can uniquely determine (locally) a solution

$$Y_0 = \phi(X_0,t) \tag{8}$$

of the algebraic system $g = 0$ and there remains an mth order system

$$\dot{X}_0 = f(X_0,\phi(X_0,t),t,0) \tag{9}$$

which has an m-manifold of smooth solutions parameterized, say, by the initial values of its components. Linear equations for higher order terms in the outer expansion can be solved successively. The asymptotic sumsof these formal expansions result in an m-manifold of smooth solutions for the full singularly perturbed system (4). This theory can be considered to be the basis for the SAPS (Smooth Approximate Particular Solution) method of Dahlquist (1969), much of the theory of Miranker (1981), and of the bounded derivative technique of Kreiss (1979). For numerical purposes, it is not usually necessary to generate many terms in such power series expansions. Indeed, the leading terms generally suffice. Thus, it is critical to note that the nth order system for X_0 is non-stiff, and thereby computationally tractable compared to that for the original stiff (m+n)th order system (4).

In addition to the smooth solutions of (4), there is also an m-manifold of rapidly decaying solutions of the nonlinear system

$$\frac{d}{d\tau}\binom{\xi}{\eta} = \binom{f}{g}(X + \varepsilon\xi, Y + \eta, \varepsilon\tau, \varepsilon) - \binom{f}{g}(X, Y, \varepsilon\tau, \varepsilon) . \tag{10}$$

Its leading terms satisfy the nonlinear system

$$\frac{d}{d\tau}\binom{\xi_0}{\eta_0} = \binom{f}{g}(X_0(0), Y_0(0) + \eta_0, 0, 0) - \binom{f}{g}(X_0(0), Y_0(0), 0, 0) ,$$

so integration determines

$$\xi_0(\tau) = - \int_\tau^\infty \frac{d\xi_0}{d\tau}\, d\tau \tag{11}$$

as a function of η_0, while η_0 must be a solution of the initial layer system

$$\frac{d\eta_0}{d\tau} = g(X_0(0), \phi(X_0(0), 0) + \eta_0, 0, 0) \tag{12}$$

which decays to zero as $\tau \to \infty$. Under our strong stability assumption concerning the eigenvalues of g_y, however, all solutions of (12) decay monotonically (in, say, the inner product norm). Again, higher order terms in the power series expansions for $\varepsilon\xi$ and η will satisfy corresponding linearized systems and we obtain an n-manifold of decaying solutions to our system (4), parameterized by the initial values for $\eta(0,\varepsilon)$, the difference between the prescribed value $y(0)$ of the fast vector and that of the smooth outer solution vector $Y(0,\varepsilon)$. We note that with less smoothness of f and g such initial layer behavior is not guaranteed simply by the eigenvalue stability of g_y (cf. the counter-example of Kreiss and Nichols (1975)).

If we now substantially complicate matters by allowing some eigenvalues of g_y to lie in the right half plane, we can no longer expect all rapidly varying solutions of (4) to decay. Imposing, instead of initial conditions, m + n boundary conditions of the general form

$$b(x(0),y(0),x(1),y(1),\varepsilon) = 0 , \qquad (13)$$

we might anticipate nonuniform behavior as $\varepsilon \to 0$ at both endpoints and seek asymptotic solutions of the form

$$\begin{aligned} x &= X(t,\varepsilon) + \varepsilon\xi(\tau,\varepsilon) + \varepsilon\mu(\sigma,\varepsilon) \\ y &= Y(t,\varepsilon) + \eta(\tau,\varepsilon) + \nu(\sigma,\varepsilon) \end{aligned} \qquad (14)$$

where $\binom{X}{Y}$ is again a smooth outer solution, where the initial layer correction $\binom{\varepsilon\xi}{\eta}$ decays to zero as the stretched variable $\tau = t/\varepsilon$ tends to infinity, and where the terminal layer correction $\binom{\varepsilon\mu}{\nu}$ decays to zero as the reflected stretched variable

$$\sigma = (1-t)/\varepsilon \tag{15}$$

tends to infinity. Here, the outer solution is determined by the mth order differential system

$$\dot{X}_0 = f(X_0, \phi(X_0,t), t, 0) \tag{16}$$

as long as the nth order algebraic system

$$g(X_0, \phi(X_0,t), t, 0) = 0 \tag{17}$$

maintains a nonsingular Jacobian g_y. To find the initial layer correction, we must first find a decaying solution of the initial layer system

$$\frac{d\eta_0}{d\tau} = g(X_0(0), \phi(X_0(0),0) + \eta_0, 0, 0) \tag{18}$$

on $\tau > 0$, and we likewise need to obtain a decaying solution of the terminal layer system

$$\frac{d\nu_0}{d\sigma} = -\, g(X_0(1), \phi(X_0(1),1) + \nu_0, 1, 0) \tag{19}$$

on $\sigma > 0$. Higher order terms of these endpoint layer corrections will satisfy only conditionally stable linear systems, presuming the Jacobian g_y now has eigenvalues in both half planes.

We naturally impose a hyperbolicity assumption, such as requiring g_y to have a uniform hyperbolic splitting with k stable and n -k unstable eigenvalues (in, perhaps, an appropriately restricted domain). The center manifold theorem (cf. Kelley (1967)) will then guarantee that the initial layer system (18) has a k-manifold of decaying solutions, while the terminal layer system (19) has an (n-k)-manifold of decaying solutions. (For a related stability problem, see Hassard (1980).) Clearly, to obtain a solution to the boundary value problem (4)-(13) of the form (14), we must guarantee that

$$\eta_0(0) = y(0) - Y_0(0)$$

and

$$\nu_0(0) = y(1) - Y_0(1)$$

lie on these respective manifolds. Determining such manifolds first arose in Levin and Levinson (1954) (see also Levin (1957)), and more geometric descriptions have been made more recently by Fenichel (1979), Kopell (1983), Kurland (1984) and Sobolev (1984), among others. For $k = n$, it is possible to determine domains of asymptotic stability for initial value problems by appropriately combining Liapunov functions for the reduced and initial layer systems (cf. Saberi and Khalil (1984)). It does not seem known whether this approach would be an effective way of estimating domains of attraction for various limiting possibilities for boundary value problems with $k < n$. Alternative hypotheses are discussed in Vasil'eva and Butuzov (1973) and O'Malley (1980). When g is linear in y, the boundary layer systems (18) and (19) are linear, and decoupling of fast-growing and fast-decaying manifolds can be conveniently handled by Householder transformations (cf. Flaherty and O'Malley (1984) and Ascher and Weiss (1983) for numerical adaptations of this approach). For g nonlinear in y, however, the discovery of nonexistence of solutions to certain model problems halted study for some years, though further understanding has been achieved lately (cf. Coddington and Levinson (1952), Vishik and Lyusternik (1960), and Howes (1980)).

When the boundary conditions are separated, we can often identify a subset of k initial conditions and of n -k terminal conditions which are asymptotically insignificant in the $\varepsilon \to 0$ limit. We then say that the limiting solution satisfies a reduced problem, consisting of the reduced mth order equation (16) and a subset of m limiting boundary conditions. Specifying such a "cancellation law" (cf. Harris (1960)) can be of tremendous help numerically, since the nonstiff reduced problem provides the asymptotic solution away from the endpoints and numerical corrections to its solution are only needed in endpoint layers. This approach, taken by Flaherty and O'Malley (1984), cannot be beat if a cancellation law is readily available. In general, however, the limiting m + n boundary conditions

$$b\big(X_0(0),\phi(X_0(0),0) + \eta_0(0), X_0(1), \phi(X_0(1)) + \nu_0(0), 0\big) = 0 \qquad (20)$$

(and higher order terms in the expansion of (13)) link the solutions of the m-dimensional reduced system (16) and the conditionally stable n-dimensional boundary layer systems (18) and (19) in a fairly complicated manner. Nonetheless, if a unique isolated solution of this combined (leading order) boundary value problem can be obtained, a corresponding asymptotic solution is readily constructed (cf., e.g., Hoppensteadt (1971) and Vasil'eva and Butuzov (1973)). The theory relates closely to the dichotomy considerations of deHoog and Mattheij (1984), and Söderlind and Mattheij (1984) which emphasize that it is essential to the integrity of computed approximations of solutions that rapidly decaying modes be integrated in the forward direction and rapidly growing modes in the backward direction.

Now imagine that we coalesce a right and a left endpoint layer solution to create an internal transition or shock layer, just as a hyperbolic tangent function asymptotes like an exponential to ±1 at ±∞ (cf. Il'in

(1969), Hemker (1977), and Flaherty (1982)). This suggests that we should also seek solutions to singularly perturbed two point problems which feature interior jumps (i.e., regions of rapid change) connecting smooth solutions of our system. For simplicity, suppose we have two different smooth outer solutions $Y_L(t,\varepsilon)$ and $Y_R(t,\varepsilon)$ of the system

$$\varepsilon \frac{dy}{dt} = g(y,t,\varepsilon) \ . \tag{21}$$

Such a jump at $\tilde{t}$ could be obtained by setting

$$y = Y_L(t,\varepsilon) + \mu(\kappa,\varepsilon) \tag{22}$$

where

$$\kappa = (t-\tilde{t})/\varepsilon \tag{23}$$

and

$$\mu \to 0 \text{ as } \kappa \to -\infty$$

while (24)

$$\mu \to Y_R(\tilde{t}+\varepsilon\kappa,\varepsilon) - Y_L(\tilde{t}+\varepsilon\kappa,\varepsilon) \qquad \text{as } \kappa \to \infty \ .$$

(We neglect, for this introductory discussion, the possibility of layers with thicknesses other than of order ε or of situations where a rescaling of the dependent variable is also necessary). The outer limits Y_{L0} and Y_{R0} must both satisfy the reduced system

$$g(Y_0,t,0) = 0 \tag{25}$$

on the appropriate t intervals, while the shock layer system

$$\frac{d\mu_0}{d\kappa} = g(Y_{L0}(\tilde{t}) + \mu_0,\tilde{t},0) \tag{26}$$

must apply on the "magnified" interval $-\infty < \kappa < \infty$ with the prescribed limits for $\mu_0(\pm\infty)$.

To be more specific, let's further restrict attention to second order vector systems in the special form

$$\varepsilon y'' = (F(x,y))' + h(x,y) \quad . \tag{27}$$

We note that such systems arise in gas dynamics, reaction-diffusion analysis, and elsewhere (cf. Maslov and Omel'yanov (1981), Smoller (1983), and Howes (1984)) when one seeks travelling wave solutions of conservation laws $u_t + (F(u))_x = \nu u_{xx}$ as the "viscosity" $\nu \to 0$. Such systems may be integrated once to obtain

$$\varepsilon y' = F(x,y) + \int^x h((s,y(s))ds \quad .$$

However, away from regions of rapid change, the limiting solution u will satisfy the limiting first order system

$$(F(x,u))' + h(x,u) = 0 \tag{28}$$

whose solution is specified by any vector u(c) or, perhaps, a more general set of n boundary conditions. We note that turning points (cf., Wasow (1965) and (1984)) occur, and the reduced system is singular, whenever the matrix F_y is singular. Further, for the limits u_L and u_R to be "attractive" rest points of the appropriate shock layer system (as well as nonsingular solutions of the reduced system on opposite sides of the jump point $\tilde{x}$,) certain stability conditions on the eigenvalues of F_y might be naturally imposed. Considering only transition layers at $\tilde{x}$ of $O(\varepsilon)$ thickness, the limiting transition layer correction term μ_0 would satisfy

$$\frac{d\mu_0}{d\kappa} = F(\tilde{x},u_{L0}(\tilde{x}) + \mu_0) - F(\tilde{x},u_{L0}(\tilde{x})) \tag{29}$$

for $-\infty < \kappa < \infty$. In order to reach the prescribed limit at ∞, we must have

$$F(\tilde{x},u_{R0}(\tilde{x})) = F(\tilde{x},u_{L0}(\tilde{x})) \quad . \tag{30}$$

This severe condition may be used to locate the jump point $\tilde{x}$. It suggests, however, that only one component of u would jump at an isolated root $\tilde{x}$, and corresponds to the classical Rankine-Hugoniot condition in gas dynamics (for a trivial shock speed) where the jump occurs in a direction normal to the shock front (cf. Majda (1983)). Various additional conditions, analogous to the classical entropy conditions, might be imposed to achieve stability and, perhaps, monotonicity of μ_0 throughout the shock layer. (We note, however, that spikes and other non-monotonic transition behavior are known to occur.) Very few specific vector problems have been completely analyzed, though some work by Fife (1979), Howes and O'Malley (1980), Howes (1983, 1984), O'Malley (1983) and van Harten and Vader-Burger (1984) is important and suggests further possible studies.

In the scalar case, we can achieve substantial progress via a phase-plane analysis and by numerical upwinding and artificial viscosity techniques, when appropriate monotonicity conditions apply. Let us simply consider the autonomous scalar equation

$$\varepsilon y'' + F_y(y)y' + h(y) = 0 \tag{31}$$

on $0 \le x \le 1$ with the Dirichlet boundary values y(0) and y(1) prescribed. Introducing the Liénard variable (cf. Stoker (1950))

$$z = \varepsilon y' + F(y) \,, \tag{32}$$

our equation becomes equivalent to the system

$$\begin{aligned} z' &= -\,h(y) \\ \varepsilon y' &= z - F(y) \end{aligned} \tag{33}$$

in the z-y plane. We could motivate this transformation by noting that z is expected to be smoother than the original variable y, and that our problem seeks a trajectory passing from the (horizontal) line $y = y(0)$ to the line $y = y(1)$ in unit time, i.e.

$$\int_{y(0)}^{y(1)} \frac{dz}{h(y)} = 1$$

holds with integration along the path. Pictorially, we have

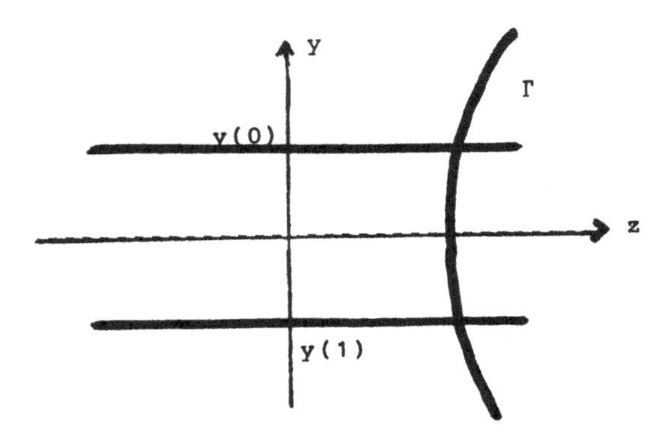

Rest points for the system occur when $h(y) = 0$ and $z = F(y)$. Thus, we can expect z to asymptotically follow the characteristic curve Γ : $z = F(y)$ unless $y' = O(1/\varepsilon)$. Motion off Γ is rapid and nearly vertical, followed by slow motion along Γ where the reduced system $F_u(u)u' + h(u) = 0$ is satisfied. For particular F and h functions, limiting solutions can be analyzed. We can, in particular, expect to study existence, uniqueness, and asymptotic behavior as $\varepsilon \to 0$ for all boundary values by examining all possible trajectories in this phase plane. Much understanding has been achieved through a geometric study of associated flows, largely led by a French-Algerian combine wanting to use non-standard analysis for singular perturbation problems (cf., especially, Lutz and Goze (1981) and Lutz and Sari (1982) as well as Callot (1981), F. Diener (1981), M. Diener (1981), (1984) and a commentary by Eckhaus (1983)).

The simplest model may be the Lagerstrom-Cole equation

$$\varepsilon y'' + yy' - y = 0$$

for which the Dirichlet problem always has a unique solution (cf. Lorenz (1983). When $-1 < y(0) < 0$ and $y(1) < 0$, for example, we have the phase portrait

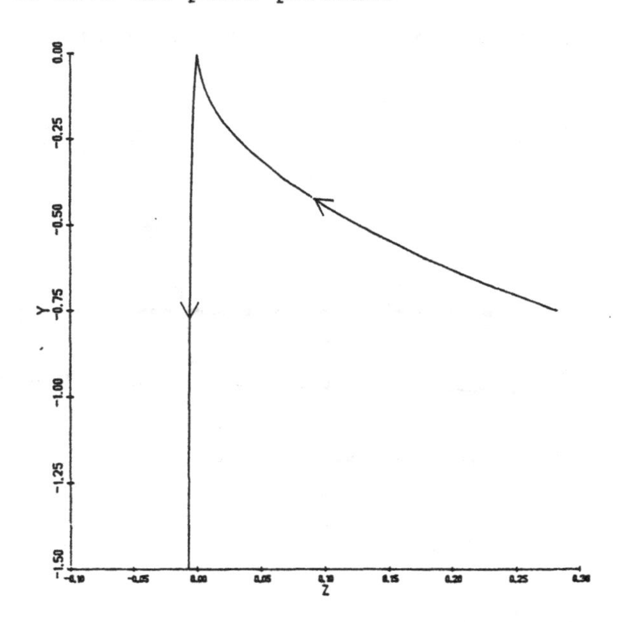

Motion slowly follows the characteristic trajectory for $-y(0)$ units of "time" x; it remains near the origin until x is nearly one, when it rapidly moves down to $y(1)$. In $y(x)$ coordinates, we then have

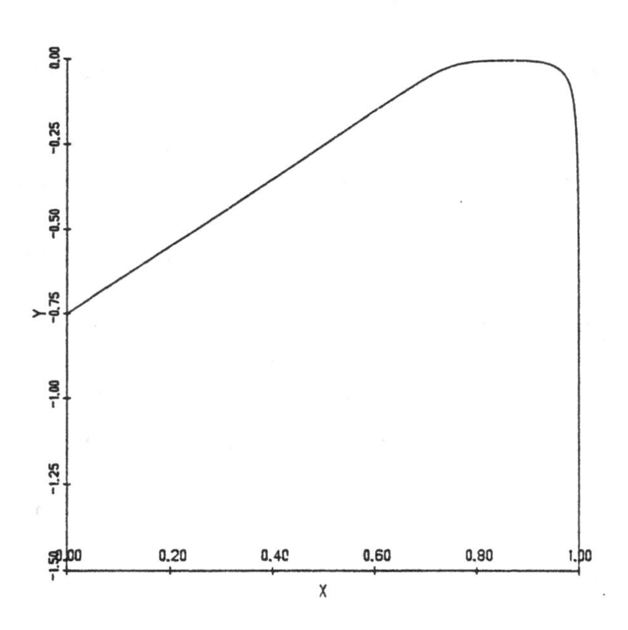

To precisely determine local behavior near $y = 0$ requires a more careful analysis, including turning point theory, but that's not necessary to obtain our crude (but perhaps adequate) idea of limiting behavior. We note that the limit $u = 0$ is actually a singular solution of the reduced

equation $uu' - u = 0$, since along it $F_u(u) = u = 0$. With other boundary values (specifically, for $-1 < y(0) + y(1) < 1$ and $y(1) > y(0) + 1$), the Cole model has an interior shock layer at $\tilde{x} = \frac{1}{2}(1-y(0)-y(1))$, as the Rankine-Hugoniot theory would predict.

Examples with multiple rest points (when, say, h has multiple zeros) provide more fascinating possibilities, and some have been studied. When h_y remains negative, we might expect some compactness arguments to give a priori solution bounds. Indeed, Lorenz (1982) shows that the Dirichlet problem for (31) has a unique solution whose limit is of bounded variation as $\varepsilon \to 0$. By using upwinding in the difference schemes, as proposed by Engquist and Osher [cf. Osher (1981)], among others, one gets excellent numerical results. Detailed turning point analysis is then unnecessary for numerical integration even if F_y has quite a few zeros. Lorenz studies the interesting example where $F(y) = \frac{y^2}{4}(2-y^2)$ and $h(y) = -y$. For end values as pictured, we obtain the limiting solution

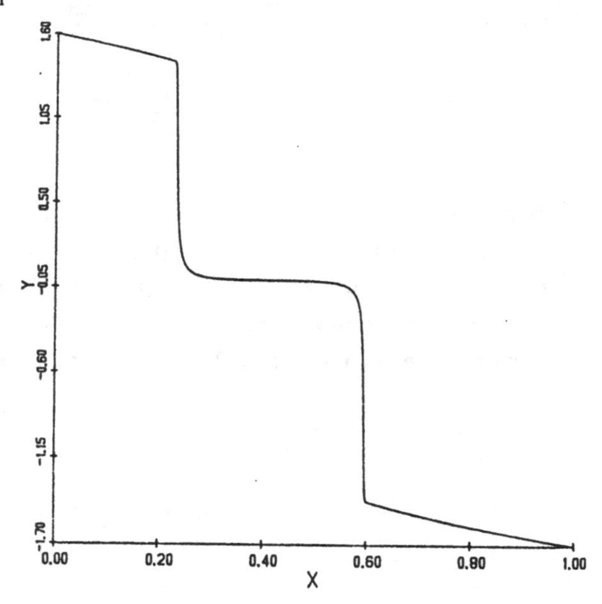

More capricious linear and nonlinear problems are included in the papers of Pearson (1968) (who used mesh refinement, continuation in ϵ, and a minimum of theory) and in the thesis of Hemker (1977) (who used exponential fitting and related finite element concepts). The need to use meshes which adapt with approximate solutions is certainly clear, but so, too, is the wisdom of using as much analytical information as can be learned about the solutions of tough, nonlinear sample problems.

Acknowledgment

The author is grateful that this research was supported in part by the National Science Foundation under Grant Number DMS-8301665 and by the U. S. Army Research Office under Grant Number DAAG29-82-K-0197.

References

[1] L. Abrahamsson and S. Osher (1982), "Monotone difference schemes for singular perturbation problems," SIAM J. Num. Anal. 19, 979-992.

[2] U. Ascher and R. Weiss (1983), "Collocation for singular perturbations I: First order systems with constant coefficients," SIAM J. Num. Anal. 20, 537-557.

[3] D. L. Brown (1982), Solution Adaptive Mesh Procedures for the Numerical Solution of Singular Perturbation Problems, Ph.D. thesis, California Institute of Technology, Pasadena.

[4] J. -L. Callot (1981), Bifurcations du Portrait de Phase pour des Équations Différentielles Linéaires du Second Ordre ayant pour Type l' Équation d' Hermite, Thèse, Université Louis Pasteur, Strasbourg.

[5] R. C. Y. Chin, G. W. Hedstrom, and F. A. Howes (1984), A Survey of Analytical and Numerical Methods for Multiple Scale Problems, preprint, Lawrence Livermore Laboratory.

[6] J. H. Chow (editor) (1982), Time-Scale Modeling of Dynamic Networks with Applications to Power Systems, Lecture Notes in Control and Information Sciences 46, Springer-Verlag, Berlin.

[7] E. A. Coddington and N. Levinson (1952), "A boundary value problem for a nonlinear differential equation with a small parameter," Proc. Amer. Math. Soc. 3, 73-81.

[8] G. Dahlquist (1969), A numerical method for some ordinary differential equations with large Lipschitz constants, Information Processing 68 (A. J. H. Morrell, editor), North-Holland, Amsterdam, 183-186.

[9] G. Dahlquist (1984), On Transformations of Graded Matrices with Applications to Stiff ODE's, preprint, Royal Institute of Technology, Stockholm.

[10] F. R. deHoog and R. M. M. Mattheij (1983), On Dichotomy and Well-Conditioning in BVP, Report, Catholic University, Nijmegen.

[11] F. Diener (1981), Méthode du Plan d'Observabilité, Thèse, Université Louis Pasteur, Strasbourg.

[12] M. Diener (1981), *Étude Générique des Canards*, Thèse, Université Louis Pasteur, Strasbourg.

[13] M. Diener (1984), "The canard unchained or how fast/slow dynamical systems bifurcate," *Math. Intelligencer* 6, no. 3, 38-49.

[14] W. Eckhaus (1983), "Relaxation oscillations, including a standard chase on French ducks," *Lecture Notes in Math.* 985, Springer Verlag, Heidelberg, 449-494.

[15] N. Fenichel (1979), "Geometric singular perturbation theory for ordinary differential equations," *J. Differential Equations* 31, 53-98.

[16] P. C. Fife (1976), "Boundary and interior transition layer phenomena for pairs of second order differential equations," *J. Math. Anal. Appl.* 54, 66-93.

[17] P. C. Fife (1979), *Mathematical Aspects of Reacting and Diffusing Systems*, Lecture Notes in Biomathematics 28, Springer Verlag, Berlin.

[18] J. E. Flaherty (1982), "A rational function approximation for the integration point in exponentially weighted finite element methods," *Int. J. Num. Meth. Eng.* 19, 782-791.

[19] J. E. Flaherty and R. E. O'Malley, Jr. (1977), "The numerical solution of boundary value problems for stiff differential equations," *Math. Comp.* 31, 66-93.

[20] J. E. Flaherty and R. E. O'Malley, Jr. (1984), "Numerical methods for stiff systems of two-point boundary value problems," *SIAM J. Sci. Stat. Comp.*

[21] P. Habets (1982), *Singular Perturbations in Nonlinear Systems and Optimal Control*, report, Université Catholique de Louvain.

[22] W. A. Harris, Jr. (1960), "Singular perturbations of two-point boundary problems for systems of ordinary differential equations," *Arch. Rational Mech. Anal.* 5, 212-225.

[23] B. D. Hassard (1980), "Computation of invariant manifolds," *New Approaches to Nonlinear Problems in Dynamics*, P. J. Holmes (editor), SIAM, Philadelphia, 27-42.

[24] P. W. Hemker (1977), *A Numerical Study of Stiff Two-Point Boundary Problems*, thesis, Mathematisch Centrum, Amsterdam.

[25] P. W. Hemker (1982), Numerical Aspects of Singular Perturbation Problems, report, Mathematisch Centrum, Amsterdam.

[26] F. Hoppensteadt (1971), "Properties of solutions of ordinary differential equations with a small parameter," Comm. Pure Appl. Math. 24, 807-840.

[27] F. A. Howes (1978), "Boundary and interior layer behavior and their interaction," Memoirs Amer. Math. Soc. 203.

[28] F. A. Howes (1980), "Some singularly perturbed superquadratic boundary value problems whose solutions exhibit boundary and shock layer behavior," Nonlinear Analysis, Theory, Meth. and Appl. 4, 683-698.

[29] F. A. Howes (1983), "Shock layer behavior in perturbed second-order systems," Proceedings, Berkeley Conference on Control and Fluid Dynamics, to appear.

[30] F. A. Howes (1983), "Nonlinear dispersive systems: Theory and examples," Studies in Appl. Math. 69, 75-97.

[31] F. A. Howes (1984), "Asymptotic structures in nonlinear dissipative and dispersive systems," Physica D, to appear.

[32] F. A. Howes (1984), "Multi-dimensional reaction-convection-diffusion equations," Proceedings, Dundee Conference on Differential Equations, to appear.

[33] F. A. Howes and R. E. O'Malley, Jr. (1980), "Singular perturbations of semilinear second-order systems," Lecture Notes in Math. 827, Springer-Verlag, Berlin, 131-150.

[34] A. M. Il'in (1969), "Differencing scheme for a differential equation with a small parameter affecting the highest derivatives," Math Notes 6, 596-602.

[35] W. G. Kelley (1984), "Boundary and interior layer phenomena for singularly perturbed systems," SIAM J. Math. Anal. 15 (1984), 635-641.

[36] N. Kopell (1983), "Invariant manifolds and the initialization problem for some atmospheric equations," preprint, Northeastern University.

[37] H. -O. Kreiss (1979), "Problems with different time scales for ordinary differential equations," SIAM J. Num. Anal. 16, 980-998.

[38] H. -O. Kreiss and N. Nichols (1975), Numerical Methods for Singular Perburbation Problems, report, Uppsala University.

[39] H. -O. Kreiss, N. K. Nichols, and D. L. Brown (1983), Numerical Methods for Stiff Two-Point Boundary Value Problems, preprint, California Institute of Technology.

[40] H. L. Kurland (1984), Singularly Perturbed Systems and the Morse-Conley Index, preprint, Boston University.

[41] M. Lentini, M. R. Osborne, and R. D. Russell (1983), The Close Relationships Between Methods for Solving Two-Point Boundary Value Problems, preprint, Simon Fraser University.

[42] J. J. Levin (1957), "The asymptotic behavior of the stable initial manifold of a system of nonlinear differential equations," Trans. Amer. Math. Soc. 85, 357-368.

[43] J. J. Levin and N. Levinson (1954), "Singular perturbations of nonlinear systems of differential equations and an associated boundary layer equation," J. Rational Mech. Anal. 3, 247-270.

[44] J. Lorenz (1981), Nonlinear Singular Perturbation Problems and the Engquist-Osher Difference Scheme, Report, Catholic University, Nijmegen.

[45] J. Lorenz (1982), "Nonlinear boundary value problems with turning points and properties of difference schemes," Lecture Notes in Math. 942, Springer Verlag, Berlin, 150-169.

[46] J. Lorenz (1983), Stability and Monotonicity Properties of Stiff Quasilinear Boundary Problems, preprint.

[47] J. Lorenz (1983), Analysis of Difference Schemes for a Stationary Shock Problem, preprint, Universitat Trier.

[48] R. Lutz and M. Goze (1981), Non-Standard Analysis, Lecture Notes in Math. 881, Springer-Verlag, Heidelberg.

[49] R. Lutz and T. Sari (1982), "Application of nonstandard analysis to boundary value problems in singular perturbation theory," Lecture Notes in Math. 942, Springer-Verlag, Berlin, 113-135.

[50] A. Majda (1983), "The stability of multi-dimensional shock fronts," Memoirs Amer. Math. Soc. 41, no. 275.

[51] Mao Zu-fan (1982), Partitioning a Stiff Ordinary Differential Equation by a Scaling Technique, report, Royal Institute of Technology, Stockholm.

[52] V. P. Maslov and G. A. Omel'yanov (1981), "Asymptotic soliton-form solutions of equations with small dispersion," Russian Math. Surveys 36:3, 73-149.

[53] R. M. M. Mattheij (1984), Decoupling and Stability of Algorithms for Boundary Value Problems, preprint, Catholic University, Nijmegen.

[54] R. M. M. Mattheij and R. E. O'Malley, Jr. (1984), "On solving boundary value problems for multi-scale systems using asymptotic approximations and multiple shooting," BIT, to appear.

[55] W. L. Miranker (1981), Numerical Methods for Stiff Equations, Reidel, Dordrecht.

[56] K. Nipp (1980), An Algorithmic Approach to Singular Perturbation Problems in Ordinary Differential Equations with an Application to the Belousov-Zhabotinskii Reaction, dissertation, Eidgen. Tech. Hochschule, Zurich.

[57] R. E. O'Malley, Jr. (1974), Introduction to Singular Perturbations, Academic Press, New York.

[58] R. E. O'Malley, Jr. (1980), "On multiple solutions of singularly perturbed systems in the conditionally stable case," Singular Perturbations and Asymptotics (R. E. Meyer and S. V. Parter, editors), Academic Press, New York, 87-108.

[59] R. E. O'Malley, Jr. (1983), "Shock and transition layers for singularly perturbed second-order vector systems," SIAM J. Appl. Math. 43, 935-943.

[60] S. Osher (1981), "Nonlinear singular perturbation problems and one-sided difference schemes," SIAM J. Num. Anal. 18, 129-144.

[61] C. E. Pearson (1968), "On a differential equation of boundary layer type," J. Math. and Physics 47, 134-154.

[62] C. E. Pearson (1968), "On non-linear ordinary differential equations of boundary layer type," J. Math. and Physics 47, 351-358.

[63] G. Peponides, P. V. Kokotovic, and J. H. Chow (1982), "Singular perturbations and time scales in non-linear models of power systems," IEEE Trans. Circuits and Systems 29, 758-767.

[64] A. Saberi and H. Khalil (1984), "Quadratic-type Liapunov functions for singularly perturbed systems," IEEE Trans. Automatic Control 29, 542-550.

[65] V. R. Saksena, J. O'Reilly, and P. V. Kokotovic (1984), "Singular perturbations and time-scale methods in control theory: Survey 1976-1983," Automatica 20, 273-293.

[66] J. Smoller (1982), Shock Waves and Reaction-Diffusion Equations, Springer-Verlag, New York.

[67] V. A. Sobolev (1984), Integral Manifolds and Decomposition of Singularly Perturbed Systems, preprint, Kuibyshev State University, USSR.

[68] G. Söderlind and R. M. M. Mattheij (1984), "Stability and asymptotic estimates in nonautonomous linear differential equations," SIAM J. Math. Anal. to appear.

[69] J. J. Stoker (1950), Nonlinear Oscillations, Wiley, New York.

[70] E. Urlacher (1980), "Equations Differentielles du Type $\varepsilon x''+f(x')+x = 0$ avec ε Petit," report, Université Louis Pasteur, Strasbourg.

[71] M. Van Dyke (1964), Perturbation Methods in Fluid Dynamics, Academic Press, New York.

[72] A. van Harten and E. Vader-Burger (1984), Approximate Green Functions as a Tool to Prove Correctness of a Formal Approximation in a Model of Competing and Diffusing Species, report, Utrecht University.

[73] A. B. Vasil'eva and V. F. Butuzov (1973), Asymptotic Expansions of Solutions of Singularly Perturbed Equations, Nauka, Moscow.

[74] M. I. Vishik and L. A. Lyusternik (1960), "Initial jump for nonlinear differential equations containing a small parameter," Soviet Math. Dokl. 1, 749-752.

[75] W. Wasow (1965), Asymptotic Expansions for Ordinary Differential Equations, Wiley, New York.

[76] W. Wasow (1970), "The capriciousness of singular perturbations," Nieuw Archief v. Wisk. 18, 190-210.

[77] W. Wasow (1984), Lectures on Linear Turning Point Theory, manuscript, University of Wisconsin, Madison.

[78] R. Weiss (1984), "An analysis of the box and trapezoidal schemes for linear singularly perturbed boundary value problems," Math. Comp. 42, 41-67.

Progress in Scientific Computing, Vol. 5
Numerical Boundary Value ODEs

Two Families of Symmetric Difference Schemes for Singular Perturbation Problems

Uri Ascher

Abstract

Singularly perturbed boundary value ordinary differential problems are considered. For numerical approximation, families of symmetric difference schemes, which are equivalent to certain collocation schemes based on Gauss and Lobatto points, are used. The performance of these two families of schemes is compared. While Lobatto schemes are more accurate for some classes of problems, Gauss schemes are more stable in general.

1. Introduction

Most of the literature available on the numerical solution of singularly perturbed problems deals with special methods. Such methods are designed for special applications or model problems, often with particular design criteria in mind (such as resolving outer solutions without resolving inner ones). The variety of special methods proposed is one indication of the difficulty (and importance) of the task involved. Indeed, most of the currently available software for boundary value ordinary differential problems (BVPs) - particularly software which is based on initial value techniques - performs poorly for "stiff" BVPs.

At the same time, the idea of having a general purpose code for such problems is rather attractive for a potential user, especially when the alternative is a special method, not yet implemented and tested, and not always precisely suited for the particular application. Moreover, experience with the code COLSYS [2] has been quite positive in general, and many practical, stiff problems have been satisfactorily solved using it. On the other hand, the code does not always perform in the best possible way, and the theory on which it is based holds strictly for nonstiff problems.

The numerical method implemented in COLSYS is collocation at Gaussian points, which gives a family of symmetric difference schemes. Thus we would like to understand better the properties of symmetric difference schemes, when applied to singularly perturbed problems. We consider two families of such schemes, based on Gauss points and on Lobatto points [3].

The basic reason for being interested in symmetric schemes is their invariance with respect to the direction of integration. This makes them robust and allows "lifting up" many results regarding initial value problems, to BVPs. However, the symmetry also means that the growth factor approximating a fast decaying exponential is merely bounded by 1 and this yields the need for a careful analysis. It turns out that some unexpected additional conditions are needed to insure stability of the algorithm, see (21) and (27) below. The schemes and their basic properties are presented in Sections 2, 3.

In Sections 4-7 we describe the performance of our two families of difference schemes on four classes of linear BVPs. In all cases the solution may have boundary layers, but varies smoothly away from the boundaries. Correspondingly, we shall construct a fine mesh near boundaries and a coarse one away, see fig. 1 in §3.

The class of BVPs considered in §4 contains only problems with fast (increasing and decreasing) solution components. This class is extended in §5, where we allow also slow solution components to be present, but only with a weak coupling between fast and slow components. An analysis of our schemes for such problems has been carried out in [3-5]. Here we describe the results in some detail, generalizing in (21) the assumption made in [3-5] under which stability is shown. It turns out that Gauss and Lobatto schemes have similar stability properties, but while Gauss schemes lose their superconvergence property, Lobatto schemes do not. Therefore, Lobatto schemes are preferable to the corresponding Gauss schemes for these classes of BVPs.

The situation changes in §6, where a wider class of problems is treated. These are singular singularly perturbed problems, see [1, 11]. Thus, both fast and slow solution components are present, with no restriction on the coupling between them. Here Lobatto schemes, while still retaining superconvergence order (which Gauss schemes do not have), are slightly less stable, as described in (28a) below. So, some tradeoff in desirable properties takes place: Some additional mesh points are needed to get rid of the instability effects and retain the full convergence order for Lobatto schemes (see (29)). Moreover, for nonlinear problems (not covered in this article), there may be additional negative effects; this is an open question (cf Schmeiser and Weiss [11]). The analysis reported in §6 is given in [1], with the exception of assumption (27) which is more general here.

In §7 we treat a particular model problem, rather than a class of problems as in previous sections. This is a Dirichlet problem for a homogeneous, second order differential equation which has a turning point, through which the solution is smooth. The stable directions of integration are from the boundaries towards the turning point. An example like this was used in Kreiss, Nichols and Brown [9]. Because the large eigenvalue changes significantly in size while integrating towards the turning point, an A-stability property of the numerical scheme (which both Gauss and Lobatto schemes possess) may not suffice. Indeed, it turns out that the algebraic stability property, which Gauss schemes possess and Lobatto schemes do not, distinguishes the former. We obtain a stability result for the midpoint scheme which we cannot establish for the trapezoidal scheme and show by a numerical example the superiority of Gauss schemes for this case.

The conclusion is that while Lobatto schemes are slightly more accurate for some classes of problems, Gauss schemes are slightly more stable in general.

2. The numerical schemes

Consider a system of n ODEs

$$\varepsilon y' = f(t,y) \qquad 0 \le t \le 1 \tag{1}$$

under the boundary conditions

$$B_0 y(0) + B_1 y(1) = b \tag{2}$$

where ε is a parameter, $0 < \varepsilon << 1$, and f may depend on ε as well.

For a numerical solution, consider a discretization by a <u>k-stage Runge-Kutta</u> scheme (Butcher [7])

$$\begin{array}{c|ccc} \rho_1 & \alpha_{11} & \cdots & \alpha_{1k} \\ \vdots & \vdots & & \vdots \\ \rho_k & \alpha_{k1} & \cdots & \alpha_{kk} \\ \hline & \beta_1 & \cdots & \beta_k \end{array}$$

where $0 \le \rho_1 \le \ldots \le \rho_k \le 1$ and $\sum_{\ell=1}^{k} \alpha_{j\ell} = \rho_j$, $\quad 1 \le j \le k$.
Thus, on a given mesh

$$\begin{aligned} \Delta&: 0 = t_1 < t_2 < \ldots < t_N < t_{N+1} = 1 \\ h_i &= t_{i+1} - t_i \qquad h = \max\{h_i;\ 1 \le i \le N\} \end{aligned} \tag{3}$$

an approximating mesh function $y_i \sim y(t_i)$, $y_{ij} \sim y(t_{ij})$, with $t_{ij} = t_i + h_i\rho_j$, is given by

$$\varepsilon h_i^{-1}(y_{i+1} - y_i) = \sum_{\ell=1}^{k} \beta_\ell f(t_{i\ell}, y_{i\ell}) \tag{4a}$$

$$\varepsilon h_i^{-1}(y_{ij} - y_i) = \sum_{\ell=1}^{k} \alpha_{j\ell} f(t_{i\ell}, y_{i\ell}) \qquad 1 \le j \le k, \quad 1 \le i \le N \tag{4b}$$

$$B_0 y_1 + B_1 y_{N+1} = b \tag{4c}$$

Other forms of presenting the scheme (4a) - (4b) are possible as well. Particularly popular is the use of $K_{i\ell} = f(t_{i\ell}, y_{i\ell})$ as the local unknowns, in place of $y_{i\ell}$, $\ell = 1,\ldots,k$.

Let us recall next piecewise polynomial collocation. Thus, let $y_\Delta(t)$ be a continuous piecewise polynomial on $[0,1]$, which reduces to a polynomial of degree at most k (denote this by $y_\Delta \in P_{k+1}$) on each subinterval $[t_i, t_{i+1}]$, and require that $y_\Delta(t)$ satisfies the boundary conditions (2) and that it satisfies the ODE (1) at the collocation points t_{ij}, $1 \le j \le k$, $1 \le i \le N$. It is well-known (see, e.g. [3]) that such a method reduces to a Runge-Kutta method (4) (with $y_\Delta(t_{ij}) = y_{ij}$, $y_\Delta(t_i) = y_i$). In particular, the formulation (4) is <u>equivalent</u> to a piecewise polynomial collocation method when $\rho_1 < \ldots < \rho_k$ if

$$\sum_{\ell=1}^{k} \beta_\ell p(\rho_\ell) = \int_0^1 p(s)ds \tag{5}$$

$$\sum_{\ell=1}^{k} \alpha_{j\ell} p(\rho_\ell) = \int_0^{\rho_j} p(s)ds \qquad j = 1,\ldots,k \tag{6}$$

for any polynomial $p \in P_k$. Thus, the coefficients of the k-stage Runge-Kutta scheme define quadrature schemes with precision (at least) k.

Our two families of schemes are based on collocation at Gauss and Lobatto points, see [3] and references therein.

I. Gauss points

The points ρ_j are zeros of the shifted Legendre polynomial. Thus, (5) holds for all $p \in P_{2k}$ and $\rho_1 > 0$, $\rho_k < 1$ (so mesh points t_i are not collocation points). For k = 1 we have the midpoint (or box) scheme

$$\varepsilon h_i^{-1} (y_{i+1} - y_i) = f(t_{i+\frac{1}{2}}, \tfrac{1}{2}(y_i + y_{i+1})) \qquad 1 \le i \le N \tag{7}$$

$(t_{i+\frac{1}{2}} = t_i + \frac{1}{2}h_i)$.

II. Lobatto points

Here $\rho_1 = 0$, $\rho_k = 1$ (so mesh points are collocation points and $\alpha_{1\ell} = 0$, $\ell = 1,\dots,k$) and (5) holds for all $p \in P_{2(k-1)}$. For k = 2 we have the trapezoidal scheme

$$\varepsilon h_i^{-1}(y_{i+1} - y_i) = \tfrac{1}{2}(f(t_i, y_i) + f(t_{i+1}, y_{i+1})) \quad 1 \le i \le N \tag{8}$$

3. Scheme properties and symmetry

It turns out that, when comparing the Gauss and Lobatto schemes, the k-stage Gauss scheme should be compared to the (k+1)-stage Lobatto scheme. This is because the cost of implementation of these two schemes is roughly the same [4]. Moreover, both schemes share the following properties:

1. When $\varepsilon >> h$, under appropriate smoothness assumptions

   $$|y(t_i) - y_i| = O(h^{2k}) \qquad 1 \le i \le N+1 \tag{9}$$

   This high order of convergence is referred to in the literature as superconvergence. For the midpoint and trapezoidal schemes we obtain second order accuracy.
2. The scheme is symmetric. This means that if the direction of integration is reversed, from t to 1-t, then the same scheme is obtained.
3. For the "test equation"

   $$\varepsilon y' = \lambda y \tag{10a}$$

   the difference scheme is

   $$y_{i+1} = \gamma\left(\frac{\lambda h_i}{\varepsilon}\right) y_i \tag{10b}$$

   with $\gamma(\zeta)$ a diagonal Pade approximation to e^ζ. We have (see [3] and references therein)

   $$|\gamma(\zeta)| \le 1 \qquad \text{re}(\zeta) \le 0 \tag{11a}$$

   i.e. all our schemes are A-stable. Also, from (3.7) of [3]

$$\gamma(\zeta) = (-1)^k \left(1 + 2k(k+1)\zeta^{-1} + O(\zeta^{-2})\right) \underset{\zeta\to-\infty}{\to} (-1)^k \tag{11b}$$

However, not all properties are shared alike by the k-stage Gauss and the (k+1)-stage Lobatto schemes. If we want to extend A-stability to problems with variable coefficients, i.e. we require

$$|y_{i+1}| \le |y_i| \tag{12a}$$

from the approximation of

$$\varepsilon y' = \lambda(t)y \qquad\qquad re(\lambda(t)) \le 0 \tag{12b}$$

then we find that Gauss schemes satisfy (12a) (because they are algebraically stable, see Burrage and Butcher [6]) while Lobatto schemes do not! For instance, the midpoint scheme gives

$$y_{i+1} = \frac{1 + \frac{h}{2\varepsilon}\lambda_{i+\frac{1}{2}}}{1 - \frac{h}{2\varepsilon}\lambda_{i+\frac{1}{2}}}\; y_i$$

and clearly

$$\left| \frac{1 + \frac{h}{2\varepsilon}\lambda_{i+\frac{1}{2}}}{1 - \frac{h}{2\varepsilon}\lambda_{i+\frac{1}{2}}} \right| \le 1 \qquad\qquad re(\lambda_{i+\frac{1}{2}}) \le 0$$

But for the trapezoidal scheme we obtain

$$y_{i+1} = \frac{1 + \frac{h}{2\varepsilon}\lambda_i}{1 - \frac{h}{2\varepsilon}\lambda_{i+1}}\; y_i \approx -\frac{\lambda_i}{\lambda_{i+1}}\; y_i \approx \ldots \approx (-1)^i \frac{\lambda_1}{\lambda_{i+1}}\, y_1$$

when $|\frac{\lambda h}{\varepsilon}| \to \infty$. So, if $|\lambda(t)| << |\lambda(0)|$ we have growth in the numerical solution. This will become particularly relevant in §7.

Let us reflect for a moment on (11b). While $e^\zeta \to 0$ as $\zeta \to -\infty$, its approximation $\gamma(\zeta)$ approaches 1 in magnitude. This is an inevitable consequence of using a symmetric scheme. It implies that

(a) we must have a fine mesh where the solution varies fast, i.e. we may ask only for a uniform approximation everywhere (and cannot skip, e.g., details of an inner solution).

(b) a careful analysis is needed to assess the applicability of these schemes in general, because fast fundamental solution components are not approximated well where the solution is smooth.

In contrast to symmetric schemes, one-sided schemes like the backward Euler scheme do satisfy $\gamma(\zeta) \to 0$ as $\zeta \to -\infty$. Thus, in certain applications, details of inner (layer) solutions can be skipped with a coarse mesh. However, these schemes appear to be less suitable for general purposes, see [5].

We continue with our two families of symmetric schemes.

We now restrict the boundary value problems under consideration as follows.

(a) The problem (1),(2) is assumed to be linear,

$$\varepsilon y' = A(t)y + q(t) \equiv f(t,y) \qquad 0 \le t \le 1 \tag{13}$$

Here again A and q may depend on ε as well.

(b) The problem is assumed to have smooth coefficients, be well-posed uniformly in $0 < \varepsilon < \varepsilon_0$ and to have a smooth solution away from boundaries. In particular, assume that there are possibly $O(\varepsilon \ell n \varepsilon)$ boundary layers, but no interior layers.

Thus, an appropriate mesh to describe the solution curve by a piecewise polynomial approximant is fine near the boundaries and coarse in between. As in [1],[3-5] we subdivide the interval [0,1] to 3 subintervals (see fig. 1):

I. On $[0, T_0\varepsilon]$ $\qquad h_i \le c\varepsilon \qquad 1 \le i \le \underline{i} \qquad (t_{\underline{i}} = T_0\varepsilon)$

II. On $[T_0\varepsilon,\ 1 - T_1\varepsilon]$ we have a coarse mesh, assumed for convenience to be uniform with step h satisfying

$$\varepsilon h^{-2} \le \kappa < \infty\ , \qquad h_i = h \qquad \underline{i} \le i < \bar{i} \tag{14}$$

III. On $[1 - T_1\varepsilon, 1]$ $\qquad h_i \le c\varepsilon \qquad \bar{i} \le i \le N \qquad (t_{\bar{i}} = 1 - T_1\varepsilon)$.

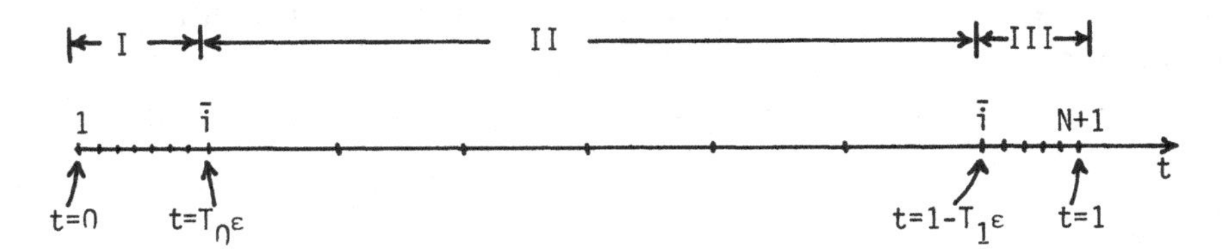

Figure 1. the mesh

4. Problems with only fast components

Assume that there exists a smooth matrix function E(t) satisfying $||E||\ ||E^{-1}|| \le$ const uniformly in t, such that

$$A(t) = E(t)\Lambda(t)E^{-1}(t)$$

$$\Lambda(t) = \begin{pmatrix} \Lambda^-(t) & 0 \\ 0 & \Lambda^+(t) \end{pmatrix} \qquad |\Lambda^-|, |\Lambda^+| \le \text{const}$$

where $\Lambda^-(t)$ is an $n_- x\ n_-$ block with eigenvalues which satisfy $re(\lambda(t)) < 0$, while $\Lambda^+(t)$ is an $n_+ x\ n_+$ block $(n_- + n_+ = n)$ with eigenvalues which satisfy $re(\lambda(t)) > 0$, $\qquad 0 \le t \le 1$.

Let

$$u(t) = E^{-1}(t)y(t) \qquad u \equiv \begin{pmatrix} u^- \\ u^+ \end{pmatrix} \equiv \begin{pmatrix} P_-E^{-1}y \\ P_+E^{-1}y \end{pmatrix} \tag{15}$$

where $P_- \in R^{n_- \times n}$ and $P_+ \in R^{n_+ \times n}$ are obvious projection matrices.

Now, for <u>analysis</u> purposes, we may consider three discretized boundary value problems separately.

I. On $[0, T_0\varepsilon]$ the difference equations (4a,b) for (13) are considered for $1 \le i < \underline{i}$ under the BC

$$u_1^- , \quad u_{\underline{i}}^+ \text{ given}$$

II. On $[T_0\varepsilon, 1-T_1\varepsilon]$ the same difference equations are considered for $\underline{i} \le i < \bar{i}$ under the BC

$$u_{\underline{i}}^- , \quad u_{\bar{i}}^+ \text{ given}$$

III. On $[1-T_1\varepsilon, 1]$ the same difference equations are considered for $\bar{i} \le i \le N$ under the BC

$$u_{\bar{i}}^- , \quad u_{N+1}^+ \text{ given}$$

If we can show stability for each of these subproblems then, matching the solutions at $T_0\varepsilon$ and at $1-T_1\varepsilon$ and requiring satisfaction of the BC (4c) yields similar stability results for the actual discretization (4) because of the assumed well-posedness of the differential problem (13),(2) (cf [13, 3-5, 1]).

Let us consider a k-stage Gauss scheme or a (k+1)-stage Lobatto scheme.

I. <u>On $[0, T_0\varepsilon]$</u>

This is a short interval and we use a fine mesh with $h_i \le c\varepsilon$. The obvious stretching transformation $\tau = {}^t/_\varepsilon$ then makes the usual infinite interval analysis applicable. We construct a mesh (Ascher and Weiss [3,4], Markowich and Ringhofer [10]) to equidistribute the error. Thus for a given error tolerance δ, choose

$$h_1 \approx \frac{\varepsilon}{|\lambda|}\delta^{\frac{1}{2k}} ; \qquad t_{i+1} := t_i + h_i \tag{16a}$$

$$h_i := h_{i-1} \exp\{\frac{1}{2k} \ \frac{-re(\lambda)}{\varepsilon} \ h_{i-1}\} \qquad i = 2,3,... \tag{16b}$$

until

$$t_{i+1} \ge \frac{\varepsilon \ \ell n\delta}{re(\lambda)} =: T_0\varepsilon \tag{16c}$$

In (16a) $|\lambda|$ is the largest magnitude of the eigenvalues of $\Lambda^-(0)$

while in (16b,c), $-\mathrm{re}(\lambda)$ is the smallest real part of these eigenvalues.

Note that T_0 is defined in (16c) so that $|u^-(T_0\varepsilon)| = \delta$. Thus the error in $[0, T_0\varepsilon]$ is $O(\delta)$ and $|u_i| = O(\delta)$. The resulting number of mesh points is proportional to $\delta^{-\frac{1}{2k}}$ and is independent of ε, see [4].

Remark: The assumption that $\mathrm{re}(\lambda(t)) < 0$, $0 \le t \le 1$, has to be replaced in practice by a requirement that

$$|\mathrm{im}(\lambda(t))| \le c|\mathrm{re}(\lambda(t))|$$

for some constant c of moderate size, for all eigenvalues of A. If an eigenvalue is very close to the imaginary axis, say $\lambda = -\varepsilon + i$, then the mesh selection (16) would simply produce too many mesh points. Other approaches should be considered in this stiff oscillatory case.

A similar treatment with similar results is given to the other short interval $[1 - T_1\varepsilon, 1]$.

II. On $[T_0\varepsilon, 1 - T_1\varepsilon]$

Here the mesh is coarse, so more surprises may occur. To indicate how difficulty may arise, consider again the test equation (10a) and its approximation (10b). Let

$$\sigma := \varepsilon h^{-2} \le \kappa \tag{17}$$

Then (11b) yields

$$y_{i+1} = (-1)^k(1 + 2k(k+1)\sigma\lambda^{-1}h + O(h^2))y_i \tag{18}$$

so, for

$$\hat{y}_i := (-1)^{ik}y_i \tag{19}$$

(18) approximates, for small h (with σ fixed), the differential equation

$$\hat{y}' = 2k(k+1)\sigma\lambda^{-1}\hat{y}$$

Some tedious, but straightforward, algebra yields a similar result for the more general case (13). Thus, the homogeneous difference equations approximate the system

$$\hat{y}' = 2k(k+1)\sigma A^{-1}\hat{y} \tag{20a}$$

So, if (20a) subject to the boundary conditions

$$P_-E^{-1}(0)\hat{y}(0) = 0, \qquad P_+E^{-1}(1)\hat{y}(1) = 0 \tag{20b}$$

has a nontrivial solution (i.e. σ is an eigenvalue of (20)) then we cannot hope for stability. We must assume that this does not occur:

Assumption 21. The boundary value problem (20) has only the trivial solution for ε and h small enough.

When κ in (17) is small enough, this assumption reduces to that made in [13] and in [4] that

$$\begin{pmatrix} P_-E^{-1}(0) \\ P_+E^{-1}(1) \end{pmatrix} \quad \text{is nonsingular} \tag{22}$$

An assumption like (21) was first made by Kreiss [8] for another centered difference scheme. The BVP (20) was first investigated for k = 1 by de Hoog and Weiss (private communication).

The following convergence estimates were obtained in [4] under assumption (22): The error at mesh points satisfies

$$|y_i - y(t_i)| \le \text{const}\{|u_{\underline{i}}^- - u^-(T_0\varepsilon)| + |u_{\bar{i}}^+ - u^+(1 - T_1\varepsilon)| + e\} \tag{23a}$$

where the first two terms propagate the error from the boundary layers and satisfy

$$|u_{\underline{i}}^- - u^-(T_0\varepsilon)| \;,\; |u_{\bar{i}}^+ - u^+(1 - T_1\varepsilon)| = O(\delta). \tag{23b}$$

The remaining error term e is given as follows: For a k-stage Gauss scheme

$$e = \begin{cases} O(h^{k+1}) & k \;\; \text{odd} \\ O(h^k) & k \;\; \text{even} \end{cases} \tag{23c}$$

while for a (k+1)-stage Lobatto scheme, the preferred estimate is

$$e = \begin{cases} \varepsilon O(h^{k+1}) & k+1 \;\; \text{odd} \\ \varepsilon O(h^k) & k+1 \;\; \text{even} \end{cases} \tag{23d}$$

Example 1 In (13), (2) let

$$A(t) = \begin{pmatrix} -\theta\sin 2t & -(1 + \theta\cos 2t) \\ 1 - \theta\cos 2t & \theta\sin 2t \end{pmatrix} \quad q(t) = \varepsilon(A^{-1}g)' - g$$

$$B_0 = \begin{pmatrix} \nu & 1 \\ 0 & 0 \end{pmatrix} \qquad B_1 = \begin{pmatrix} 0 & 0 \\ \nu\cos 1 - \sin 1 & -\nu\sin 1 - \cos 1 \end{pmatrix} \quad b = \begin{pmatrix} 0 \\ 0 \end{pmatrix}$$

where

$$\theta = \mu(\mu^2 - 1)^{-\frac{1}{2}}, \; \nu = \mu - \sqrt{\mu^2 - 1}, \; \mu = \pi/2, \; g(t) = \begin{pmatrix} \sin\pi t \\ \sin\pi t \end{pmatrix} .$$

The exact solution is

$$y(t) = A^{-1}g$$

(i.e. there are no boundary layers because of the particular choice of the boundary values), the eigenvalues of A are $\pm(\mu^2 - 1)^{-1/2}$, the problem is well conditioned and the auxiliary problem (20) has nontrivial solutions when $\sigma = \frac{1}{2k(k+1)}$. Thus, assumption (21) is violated for such combinations of h and ε.

In table 1 we list some numerical results. The meshes were constructed with a uniform step size h, superimposed on layer meshes as in (16), with $\delta = .01$. By Gk we denote a k-stage Gauss scheme and by Lk, a k-stage Lobatto scheme. e is the maximum error in the first solution component over the uniform part of the mesh and .a - b means .a x 10^{-b}. Note that when specifying h and σ, $\varepsilon = \sigma h^2$ is specified as well.

Table 1. Numerical results for example 1

σ	h	e(G1)	e(L2)	e(G2)	e(L3)	e(G3)	e(L4)
1/4	.1	.79	.17-2	.30-1	.31-3	.13-3	.27-5
1/4	.05	1.62	.48-3	.73-2	.20-4	.74-5	.44-7
1/12	.1	.25	.53-3	.46+1	.25-2	.24-3	.12-5
1/12	.05	.67-1	.34-4	.25+1	.29-3	.19-4	.19-7
1/24	.1	.24	.28-3	.1	.72-4	.27-1	.11-4
1/24	.05	.60-1	.18-4	.25-1	.45-5	.21-2	.20-6

From table 1 it is clear that the errors deteriorate when σ attains a critical value. (These errors are underlined). This is particularly noticeable for the Gauss schemes, because of the difference between (23c) and (23d).

In addition, if we now remove the layer meshes and apply the midpoint scheme with h = .1, σ = 1/4, we obtain

$$e(G1) = .89\text{-}1 \qquad (< .79)$$

So, the error using the coarser subgrid is better! This is because the mesh structure allowing for the peculiar singularity in (20) to be relevant has been violated (and, at the same time, y(t) is smooth). Perhaps this unusual calculation can also indicate that the requirement (21), which is a restriction on symmetric schemes not shared by upwinded, one-sided schemes, is mainly of theoretical interest and is almost always satisfied in practice.

5. Problems with fast and slow components

The more usual situation in problems with different time scales is that there are both fast and slow solution components. The simplest case, treated in [4], is when A(t) of (13) can be written as

$$A(t) = \begin{pmatrix} A^{11} & A^{12} \\ \varepsilon A^{21} & \varepsilon A^{22} \end{pmatrix} \qquad |A^{j\ell}| \le \text{const.} \tag{24a}$$

where A^{11} now has a regular splitting, like A in the previous section. Corresponding to the partition in A we write

$$q(t) = \begin{pmatrix} q^1 \\ \varepsilon q^2 \end{pmatrix} \qquad y = \begin{pmatrix} y^1 \\ y^2 \end{pmatrix} \tag{24b}$$

with y^1 the fast components and y^2 the slow ones.

It turns out that the treatment on the short layer intervals is similar to that of the previous section and also assumptions (21) and (22) extend directly with A^{11} replacing A, y^1 replacing y. The only significant difference is in the convergence estimates for e of (23). Here, the factor in (23d) is lost, but the full superconvergence order is retained, i.e. for the (k+1)-stage Lobatto scheme

$$e = O(h^{2k}). \tag{25}$$

For the k-stage Gauss scheme, the same loss of superconvergence order as in (23c) holds for the fast solution components, while for the slow ones the usual superconvergence order (25) is retained under some restrictions on the boundary conditions, see [4].

So, for the classes of problems considered hitherto, Lobatto schemes are generally favoured over corresponding Gauss schemes.

6. Singular singularly perturbed problems

Let us now further extend the class of problems under consideration. We assume that there exists matrix function E(t) as before, such that

$$A(t) = E(t)\ \Lambda(t)\ E^{-1}(t)$$

$$\Lambda(t) = \begin{pmatrix} \Lambda^-(t) & & 0 \\ & \Lambda+(t) & \\ 0 & & \varepsilon\Lambda^0(t) \end{pmatrix} \begin{matrix} \updownarrow n_- \\ \updownarrow n_+ \\ \updownarrow n_s \end{matrix} \quad \updownarrow n_r \qquad |\Lambda^-|,\ |\Lambda^+|,\ |\Lambda^0| \le \text{const}$$

with Λ^- and Λ^+ negative and positive blocks as in §4. The dependence of A and Λ on ε has been suppressed for clarity of notation; at $\varepsilon = 0$ A is clearly singular, with a nondefective nullspace.

Defining

$$w(t) := E^{-1}(t)y(t) \qquad w \equiv \begin{pmatrix} u \\ v \end{pmatrix} \equiv \begin{pmatrix} P_r E^{-1} y \\ P_s E^{-1} y \end{pmatrix} \quad u \equiv \begin{pmatrix} u^- \\ u^+ \end{pmatrix} \equiv \begin{pmatrix} P_- P_r E^{-1} y \\ P_+ P_r E^{-1} y \end{pmatrix} \tag{26}$$

we split w into n_r fast components u and n_s slow components v, $n_r + n_s = n$. Subsequently, u is further split as in §4 to n_- decreasing components and n_+ increasing ones.

We can now reformulate the differential equations (13) in terms of w instead of y, and the form of the resulting problem is that of §5. However, the numerical analysis of [4] still had to be extended in Ascher [1], because the difference schemes are applied to the original equations (13), not the transformed ones. A similar transformation from y to w for the numerical scheme introduces changes on the scale of h, which are potentially crucial when the problem contains the scale of ε and $\varepsilon << h$.

<u>On $[0, T_0\varepsilon]$</u> under the BC

$$u_1^-, \quad u_{\underline{i}}^+, \quad v_1 \quad \text{given}$$

an analysis yields similar results to those in §4. Analogous results are obtained on the other short interval $[1 - T_1\varepsilon, 1]$, where the mesh is fine as well.

<u>On $[T_0\varepsilon, 1 - T_1\varepsilon]$</u> under the BC

$$u_{\underline{i}}^-, \quad u_{\bar{i}}^+, \quad v_{\underline{i}} \quad \text{given}$$

the analysis is more complicated. With $\sigma = \varepsilon h^{-2}$ fixed, the analysis is performed in terms of the transformed variable w (cf. [1], [4]). The stability assumption (21) becomes

<u>Assumption 27</u> The boundary value problem

$$\hat{u}' = P_r[-E^{-1}E' + 2k(k+1)\sigma\, \Lambda^+]P_r^T \hat{u}$$

$$\hat{u}^-(0) = 0 \qquad \hat{u}^+(1) = 0$$

has only the trivial solution for ε and h small enough.

Thus, even if we set $\sigma = 0$, the condition here does not reduce to an algebraic one like (22), cf [1, eqns (2.24), (2.25)].

Next, consider the convergence estimates. It turns out that (23a,b,c) still holds for the k-stage Gauss scheme. For the (k+1)-stage Lobatto scheme, however, there is a change, due to lack of D-stability (van Veldhuisen [12]). Instead of (23a) we get

$$|y_i - y(t_i)| \le \text{const}\{\varepsilon^{-1}h^{\hat{k}}[|u_{\underline{i}}^{-} - u^{-}(T_0\varepsilon)| + |u_{\bar{i}}^{+} - u^{+}(1 - T_1\varepsilon)|] + e\} \tag{28a}$$

where, as in (25)

$$e = O(h^{2k}) \tag{28b}$$

and

$$\hat{k} = \begin{cases} k+1 & k \text{ odd} \\ k & k \text{ even} \end{cases} \tag{28c}$$

The implication of (28a) is that the errors propagating from the boundary layers are blown up by a factor $O(\varepsilon^{-1}h^{\hat{k}})$. Thus, if we want $O(\delta)$ uniform accuracy, we must make sure that the error when leaving the fine boundary layer meshes is $O(\delta\varepsilon h^{-\hat{k}})$, when $\varepsilon << h^{\hat{k}}$. This is done by extending the layer meshes. For the left end we now take

$$T_0\varepsilon = \frac{\varepsilon \ \ell n(\delta\varepsilon h^{-\hat{k}})}{re(\lambda)} \tag{29a}$$

(so $u^{-}(T_0\varepsilon) = \delta\varepsilon h^{-\hat{k}}$). The mesh is then constructed as follows. Up to $t = \frac{\varepsilon \ \ell n\delta}{re(\lambda)}$, the previous construction of (16) is used to retain a uniform $O(\delta)$ accuracy. Then we continue to add mesh points to insure that

$$|u_{\underline{i}}| \lesssim 2|u^{-}(T_0\varepsilon)|$$

This is done with a uniform step size

$$h_L = \frac{\varepsilon}{|\lambda|}|\ell n(\varepsilon^{-1}h^{\hat{k}})|^{-\frac{1}{2k}} \tag{29b}$$

see [1, (4.24) - (4.28), (4.49) - (4.50)]. This gives some more mesh points for the Lobatto scheme. Indeed, the number of mesh points mildly increases, for the first time, as ε decreases.

<u>Example 2</u> [1]. The second order problem

$$\varepsilon y'' + (2 + \cos\pi t)y' - y = f(t)$$
$$y(0) = 0, \quad y(1) = -1$$

has, with f(t) appropriately chosen, the solution

$$y(t) = \cos\pi t - e^{-3t/\varepsilon} + O(\varepsilon^2)$$

Consider two reformulations to a first order system:

(a) $\varepsilon y' = -(2+\cos\pi t)y + z$

$z = (1-\pi\sin\pi t)y \qquad + f(t)$

This reformulation is of the type discussed in §5 (and in [4]) with y as a fast component.

(b) $\varepsilon y' = \qquad z$

$\varepsilon z' = \varepsilon y - (2+\cos\pi t)z + \varepsilon f(t)$

This reformulation is of the type discussed in this section (and in [1]) with y as a rescaled slow component.

In table 2 we display results of some numerical calculations for $\varepsilon = 10^{-10}$, $\delta = 10^{-8}$. Such a small tolerance on the layer errors requires over 5700 mesh points, if the midpoint or trapezoidal schemes are to be used; hence we use higher order schemes. The notation is similar to that of table 1. For a k-stage Gauss scheme or a (k+1)-stage Lobatto scheme, we superimpose a uniform mesh with step size h on layer meshes constructed as in (16). We denote the number of mesh points of the resulting mesh by N(Gk) and the error in y away from layers by e(Gk) or e(L(k+1)). In the case of Lobatto schemes for form (b) we also add mesh points as in (29), denote the resulting total number of mesh points by N* and the error by e*.

Table 2. Numerical results for example 2 with $\varepsilon = 10^{-10}$, $\delta = 10^{-8}$

form	h	k	N(Gk)	e(Gk)	e(L(k+1))	N*(L(k+1))	e*(L(k+1))
(a)	.2	2	84	.25-1	.37-3		
	.1	2	89	.63-2	.22-4		
(b)	.2	2	84	.71-4	.57+1	93	.70-4
	.1	2	89	.44-5	.72	97	.44-5
(a)	.2	3	25	.22-2	.48-5		
	.1	3	30	.10-3	.12-6		
(b)	.2	3	25	.23-6	.10	29	.10-6
	.1	3	30	.52-7	.94-2	34	.61-8

The reader can clearly see the loss of superconvergence order for Gauss points in form (a), while Lobatto schemes retain the full order (25). Using form (b), on the other hand, shows the Gauss schemes in a better light. Since y is essentially slow here, the convergence order is high, while the Lobatto schemes display the lack of D-stability and require the additional mesh points (29) to obtain good accuracy (compare e and e*).

7. A turning point problem

In the previous section we have seen that, in the comparison between the two families of schemes, the superior superconvergence properties of Lobatto schemes are countered by better stability properties of Gauss schemes. Here we see that this trend continues by treating an example of a turning point problem

$$\varepsilon y'' = tay' + by \qquad -1 \le t \le 1 \tag{30a}$$

$$y(-1) = y_1 , \quad y(1) = y_{N+1} \tag{30b}$$

with a and b constants, $a > b > 0$. The eigenvalue $\frac{ta}{\varepsilon}$ is large and negative at $t = -1$ and large and positive near $t = 1$. This allows a stable solution construction which has boundary layers at the interval ends $t = \pm 1$. The solution is transcendentally small when approaching $t = 0$ from both sides, so a smooth matching takes place at the turning point.

We rewrite (30a) as a first order system

$$\begin{aligned} \varepsilon y' &= tay + z \\ z' &= (b-a)y \end{aligned} \qquad -1 \le t \le 1 \tag{31}$$

and use a mesh with the construction of previous sections. Thus, near $t = \pm 1$ we use (16) for a given tolerance δ and elsewhere we use a uniform mesh with step size h.

The analysis of our schemes on the layer subintervals $[-1, -1 + T_0\varepsilon]$ and $[1 - T_1\varepsilon, 1]$ is precisely the same as in [4]. However, on the long interval $[-1 + T_0\varepsilon, 1 - T_1\varepsilon]$ the situation is different. Corresponding to the stable directions of integration of (31), we consider the numerical schemes for an initial value problem on $[-1 + T_0\varepsilon, 0]$ and for a terminal value problem on $[0, 1 - T_1\varepsilon]$. For simplicity, assume that $t_M = 0$ is a mesh point and consider the <u>midpoint scheme</u>

$$\varepsilon h^{-1}(y_{i+1} - y_i) = \tfrac{1}{2} a t_{i+\frac{1}{2}}(y_i + y_{i+1}) + \tfrac{1}{2}(z_i + z_{i+1}) \tag{32a}$$

$$\underline{i} \le i \le M$$

$$h^{-1}(z_{i+1} - z_i) = \tfrac{1}{2}(b-a)(y_i + y_{i+1}) \tag{32b}$$

with $t_{\underline{i}} := -1 + T_0\varepsilon$. Thus, $t_{i+\frac{1}{2}} < 0$ but $t_{i+\frac{1}{2}}$ is not bounded away from 0 uniformly in h.

Defining

$$\hat{z}_i := h^{-1} z_i \tag{33}$$

and substituting (32b) in (32a), we have

$$\varepsilon h^{-1}(y_{i+1} - y_i) = \frac{at_{i+\frac{1}{2}}}{b-a}(\hat{z}_{i+1} - \hat{z}_i) + \frac{h}{2}(\hat{z}_{i+1} + \hat{z}_i)$$

$$\left(1 + \frac{(b-a)h}{2at_{i+\frac{1}{2}}}\right)\hat{z}_{i+1} = \left(1 - \frac{(b-a)h}{2at_{i+\frac{1}{2}}}\right)\hat{z}_i + \frac{\varepsilon h^{-1}(b-a)}{at_{i+\frac{1}{2}}}(y_{i+1} - y_i) \tag{34}$$

With

$$\zeta_i := \frac{(b-a)h}{2at_{i+\frac{1}{2}}} > 0 , \qquad \eta_i := \frac{2\varepsilon h^{-1}}{at_{i+\frac{1}{2}}} < 0$$

we can write (34) and (32a) as

$$\hat{z}_{i+1} = \frac{1 - \zeta_i}{1 + \zeta_i}\hat{z}_i + \frac{\varepsilon h^{-1}(b-a)}{at_{i+\frac{1}{2}} + \frac{1}{2}h(b-a)}(y_{i+1} - y_i) \tag{35a}$$

$$\underline{i} \le i \le M$$

$$y_{i+1} = -\frac{1 + \eta_i}{1 - \eta_i}y_i - \frac{h}{at_{i+\frac{1}{2}} - 2\varepsilon h^{-1}}(\hat{z}_i + \hat{z}_{i+1}) \tag{35b}$$

from which we obtain

$$||\hat{z}_\Delta||_- \le \text{const}\{|\hat{z}_{\underline{i}}| + \varepsilon h^{-2}||y_\Delta||_-\} \tag{36a}$$

$$||y_\Delta||_- \le \text{const}\{|y_{\underline{i}}| + ||\hat{z}_\Delta||_-\} \tag{36b}$$

where

$$||\psi_\Delta||_- := \max_{\underline{i} \le i \le M} |\psi_i| \tag{37}$$

Then, for εh^{-2} sufficiently small we obtain from (36)

$$||\hat{z}||_{\Delta-} , ||y_\Delta||_- \le \text{const}\{|\hat{z}_{\underline{i}}| + |y_{\underline{i}}|\} \tag{38a}$$

A similar stability bound is obtained for the positive subinterval, viz.

$$\max_{M \le i \le N} |y_i| , \max_{M \le i \le N} |\hat{z}_i| \le \text{const}\{|\hat{z}_{\bar{i}}| + |y_{\bar{i}}|\} \tag{38b}$$

where $t_{\bar{i}} := 1 - T_1\varepsilon$. Note that no such estimates are available for the trapezoidal scheme!

Now, if the approximations on the two subintervals are matched successfully at t_M then the error control (16), with (38), yields

$$|y_i - y(t_i)| \le \text{const}\ \delta \qquad 1 \le i \le N + 1 \tag{39}$$

and in general we cannot hope for more. A similar result can be obtained for higher order Gauss schemes, but not for Lobatto schemes.

Example 3 [9] In (30) we take $a = 1$, $b = \frac{1}{2}$, $y_1 = 1$, $y_{N+1} = 2$. The solution can be written as

$$y(t) = e^{-(t+1)/\varepsilon} + 2e^{(t-1)/\varepsilon} + O(\varepsilon).$$

As before, we superimpose a sparse uniform ($h = .2$) mesh over fine layer meshes and denote the resulting number of mesh points by N and the maximum error away from layers (including at $t = 0$) by $e(y)$. The results in table 3 clearly indicate what a lack of algebraic stability can do: whereas the errors of Gauss schemes are $\approx\delta$, those of Lobatto schemes are much larger.

Table 3. Numerical results for example 3, $\varepsilon = 10^{-6}$, $h = .2$

scheme	δ	N	e(y)	N*	e*(y)
midpoint	10^{-2}	24	.14-1		
trapezoidal	10^{-2}	24	.10+2	54	.41-3
Gauss 3	10^{-6}	28	.19-5	36	.71-15
Lobatto 4	10^{-6}	28	.59-1	36	.22-10

Note that we can further decrease the error away from layers for this homogeneous problem by extending the layer meshes, in principle in the same way as described in §6. We have used (29) with $\hat{k} = 0$, yielding errors $e^*(y)$ for mesh sizes N^* as listed in table 3. The error in Gauss schemes is still better than in Lobatto schemes, but at least now the latter is pleasantly reduced. At the same time, note that the maximum in $e(y)$ is achieved for Lobatto schemes near $t = 0$, so the reduction in $e^*(y)$ was achieved by treating nonlocal effects - a tough job for a general, adaptive code.

Acknowledgement: In attempting to understand example 3, I have benefitted from discussions with Georg Bader. Further results of a joint investigation into symmetric, algebraically stable schemes, will be reported in the near future.

References

[1] U. Ascher, "On some difference schemes for singular singularly-perturbed boundary value problems", Numerische Mathematik, to appear.

[2] U. Ascher, J. Christiansen and R.D. Russell, "Collocation software for boundary value ODEs", Trans. Math. Software 7 (1981), 209-222.

[3] U. Ascher and R. Weiss, "Collocation for singular perturbation problems I: First order systems with constant coefficients", SIAM J. Numer. Anal. 20 (1983), 537-557.

[4] ________, "Collocation for singular perturbation problems II: Linear first order systems without turning points", Math. Comp. 43 (1984), 157-187.

[5] ________, "Collocation for singular perturbation problems III: Nonlinear problems without turning points", SIAM J. Scient. Stat. Comp., to appear.

[6] K. Burrage and J. Butcher, "Stability criteria for implicit Runge-Kutta methods", SIAM J. Numer. Anal. 16 (1979), 46-57.

[7] J. Butcher, "Implicit Runge-Kutta processes", Math. Comp. 18 (1964), 50-64.

[8] H. Kreiss, "Centered difference approximation to singular systems of ODEs", Symposia Mathematica X (1972), Inst. Nazionale di Alta Mathematica.

[9] H. Kreiss, N. Nichols and D. Brown, "Numerical methods for stiff two-point boundary value problems", MRC Tech. Rep. 2599 Madison, Wis. 1983.

[10] P. Markowich and C. Ringhofer, "Collocation methods for boundary value problems on 'long' intervals", Math. Comp. 40 (1983), 123-150.

[11] C. Schmeiser and R. Weiss, "Asymptotic and numerical methods for singular singularly perturbed boundary value problems in ordinary differential equations", Proc. BAIL III, J. Miller, ed., 1984.

[12] M. van Veldhuisen, "D-stability", SIAM J. Numer. Anal. 18 (1981), 45-64.

[13] R. Weiss, "An analysis of the box and trapezoidal schemes for linear singularly perturbed boundary value problems", Math. Comp. 42 (1984), 41-68.

Progress in Scientific Computing, Vol. 5
Numerical Boundary Value ODEs

A Numerical Method for Singular Perturbation Problems with Turning Points

David L. Brown †

Applied Mathematics 217-50
California Institute of Technology
Pasadena, CA 91125 USA

0. Introduction. Until recently, the development of general numerical methods for singular perturbation problems whose solutions exhibit internal layer type behavior has been largely neglected. Indeed, even the analytic study of general systems of first order ODEs with this type of behavior appears to be quite limited; perhaps this is one of the reasons for the lack of progress in this area. In this paper, we report on results obtained with H. O. Kreiss and N. Nichols [7], which address this problem. We will give a presentation which is somewhat different than that in [7], with the hope of emphasizing the similarity of our approach to the ideas that underlie the analytic technique of matched asymptotic expansions. We also present some recent results for nonlinear problems with internal layer behavior which have been obtained together with W. L. Kath and H. O. Kreiss. All of our results are for the two-point boundary value problem for systems of first-order ordinary differential equations. Since the solutions of singular perturbation problems of this type typically vary on two or more scales, these problems are often called "stiff" boundary value problems as well.

The theory of difference approximations for the two-point boundary value problem for linear singularly perturbed ODEs *without* turning points is fairly well developed. Under appropriate assumptions, the solutions of such problems are typically smooth away from the endpoints of the interval and have possible "boundary layers" that occur in very small regions near the boundaries and connect the boundary values to the smooth interior solutions. A successful approach both theoretically and practically has been discussed by Weiss and Ascher [1], [2], [9], which is to use centered high-order two-point (implicit Runge-Kutta type) schemes for these problems together with a careful construction of the computational mesh so as to resolve the rapid boundary layer behavior. These schemes are very attractive since in contrast with the competing one-sided schemes, they do not require a (potentially expensive) change of variables before they can be stably applied. One should note, however, that error estimates do not follow directly for these schemes. It has been pointed out by both Kreiss [5], [6] and Ascher [1] that there is a special eigenvalue condition (not directly related to the original continuous problem) which must be satisfied in order that the error in the computed solution be of reasonable size. It is still not clear how serious this restriction is from a practical standpoint.

---

† present address: Los Alamos National Laboratory, Los Alamos, NM 87545 USA

For problems with turning points, the theory for difference approximations is less well developed. One would like to have a result of the type that if a mesh has been constructed which will resolve the features (e.g. internal and boundary layers) of the solution of the analytic problem, then the difference approximation that one applies on this mesh will give an accurate solution. Experimentally, one observes that with centered schemes, this is sometimes the case, and sometimes not (see e.g. [7]), while for one-sided schemes, theoretical results of this type exist (see section 4 and [7]). Regardless of which difference methods are applied, we feel that the main practical difficulty for turning point problems is one of constructing an appropriate mesh on which to solve the problem numerically. Thus the main thrust of the results discussed in sections 2 through 4 is to determine a priori where the solutions of a system of ODEs can be expected to vary rapidly.

It turns out that in order to do this, it is necessary to construct a change of variables which separates the rapidly growing, rapidly decaying, and slowly varying components of the solution. Since this is the same transformation that is required in order to apply one-sided difference approximations, we have chosen to do so in our implementation of the method.

1. Turning points. Typically, the occurence of a "turning point" in a differential equation corresponds to some kind of change of behavior in the solutions of that equation near that point. Often, the solution near such a point varies on a much different scale than it does elsewhere, and it is clear that if an accurate numerical solution of the problem is to be obtained, the computational mesh must accomodate this kind of behavior. The point of view we will take in designing our method for these problems is to detect regions with potential rapid variation a priori, and also to determine the appropriate scaling of the independent variable there, thus allowing us to construct an appropriate computational mesh. Since from a numerical point of view, the difficulties involved with turning point problems have to do not specifically with these points themselves, but with rapid variation of the solutions *nearby,* we will tend to refer to "turning point regions" instead of just "turning points". An "operational" definition of a turning point region is given below in section 3.

Before presenting the details of our method, it is instructive to consider a simple analytic example. We will look at the problem

$$\epsilon\frac{d^2y(x)}{dx^2} + 2x\frac{dy(x)}{dx} = 0, \quad -1 \leq x \leq 1 \tag{1}$$

with boundary conditions $y(-1) = 1$, $y(1) = 3$. A leading order asymptotic representation of the solution can be constructed by a trivial application of the method of matched asymptotic expansions (see e.g. Kevorkian and Cole [4].) There are three different regions of the interval $-1 \leq x \leq 1$ in each of which the behavior of the solution can be fundamentally different. Away from $x = 0$, the first term in (1) can be neglected in comparison with the second, and the solutions of (1) approximately satisfy

$$dy/dx = 0, \quad |x| > O(\epsilon). \tag{2}$$

Near $x = 0$ the two terms in (1) might easily be of similar magnitude. The appropriate "scaling" of the independent variable in this region is ϵ. Thus we introduce a new independent variable $\tilde{x}$ by $x = \epsilon \tilde{x}$, and denote $Y(\tilde{x}) = y(\epsilon \tilde{x})$, whence (1) becomes

$$\frac{d^2Y}{d\tilde{x}^2} + \tilde{x}\frac{dY}{d\tilde{x}} = 0, \quad \tilde{x} \to \pm\infty. \tag{3}$$

An approximation to the solution of this problem in the three different regions is therefore given by

$$y(x) = \begin{cases} y_L(x) = 1, & -1 \le x < 0 - O(\epsilon) \\ Y(x/\epsilon) = B\ erf(x/\epsilon) + C, & |x| < O(\epsilon) \\ y_R(x) = 3, & 0 + O(\epsilon) < x \le 1. \end{cases} \tag{4}$$

The constants B and C are determined by the "matching conditions"

$$\lim_{x \to 0-} y_L(x) = \lim_{\tilde{x} \to -\infty} Y(\tilde{x}), \quad \lim_{x \to 0-} y_R(x) = \lim_{\tilde{x} \to \infty} Y(\tilde{x}).$$

which lead to $B = 1$, $C = 2$.

The essential features of this analytic method which should be stressed are the following: Several regions with different scalings of the independent variable were determined. In each of these regions, the dominant behavior of the solution is well-described in terms of the solutions of another related differential equation. A representation of the solution is constructed from the solutions in the subregions by applying the "matching principle", which leads essentially to continuity conditions connecting those solutions.

Although our numerical method for solving turning point problems was not developed using asymptotic methods as a guideline, there are a number of interesting similarities in the two techniques. The main similarity is the underlying principle that we look for a scaling of the independent variable such that the solution of the problem will be smooth with respect to the scaled coordinates. As is discussed below, a byproduct of the automatic determination of this scaling function is that the system of ODEs is transformed to a normal form in which the "fast" variables are essentially decoupled from the "slow" variables. Since the identity of the fast and slow variables depends on the eigenstructure of the coefficient matrix, and in turning point problems this eigenstructure changes as a function of the independent variable, it is natural that we divide the interval of interest into subintervals on each of which this transformation leads to a different block structure of the coefficient matrix. For a problem in which an explicit small parameter "ϵ" can be identified (this is not necessary in the numerical case), this normal form also naturally leads to an appropriate asymptotic expansion in each of the computational subintervals and the subintervals are analogous to those discussed in the simple analytic example above. A difference is that while in the analytic case, the scalings which are determined are constant in different subintervals, the scaling function determined by our technique is essentially a continuous function. For this reason, the question of "overlap regions" and "matching" never really comes up in the numerical case; the solution and the stretching functions are both approximations to continuous functions on the entire interval.

2. Characterization of solution smoothness. Consider the two-point boundary value problem on the interval $x \in I := [0,1]$ for a system of $n = n_+ + n_- + n_o$ first order linear ODEs:

$$\frac{d}{dx}\begin{pmatrix} y(x) \\ w(x) \end{pmatrix} = \begin{pmatrix} A_{11}(x) & A_{12}(x) \\ A_{21}(x) & A_{22}(x) \end{pmatrix} \begin{pmatrix} y(x) \\ w(x) \end{pmatrix} + \begin{pmatrix} f(x) \\ g(x) \end{pmatrix} \tag{5a}$$

$$B_1 \begin{pmatrix} y(0) \\ w(0) \end{pmatrix} + B_2 \begin{pmatrix} y(1) \\ w(1) \end{pmatrix} = \gamma. \tag{5b}$$

where $y,\ f \in R^{n_+ + n_-}$, $w,\ g \in R^{n_o}$, $\gamma \in R^n$, and A_{ij}, B_i, $i,j = 1,2$ are matrices of the appropriate dimensions. Also for convenience we denote by $A(x)$ the matrix with block entries $A_{ij}(x)$, $i,j = 1,2$. We are interested in solving this problem on a (for the moment assumed) uniform mesh with meshsize which we denote by h, which we assume is fine enough to resolve any unsmoothness in the solution of the problem. (Here and elsewhere in this paper, an "unsmooth" function refers to a function whose derivatives become large.) If $h||A(x)|| << 1$ held everywhere, then we would have no difficulty in solving this problem using standard techniques. However, the class of problems we are interested in is where $h||A_{2j}|| << 1$ but $h||A_{1j}|| >> 1$. Furthermore, since we are interested in solving problems which vary on different scales in different parts of the interval, we should realize that a uniform mesh will be inappropriate for such problems, and we need to be able to find an appropriate nonuniform mesh.

The smoothness of the solutions of (5) can be characterized in terms of the smoothness of the coefficients in (5a) if the following assumption holds (note that this assumption is relaxed substantially in section 3):

ASSUMPTION: The matrix function $A(x)$ is *essentially diagonally dominated,* i.e., the elements a_{ij} of A_{11} satisfy

$$\begin{aligned} &\mathrm{Re}\,a_{ii} < 0,\ i = 1,2,\dots,n_- \\ &\mathrm{Re}\,a_{ii} > 0,\ i = n_-+1, n_-+2, \cdots, n_-+n_+ \end{aligned} \tag{6a}$$

$$\sum_{j \neq i} |a_{ij}| \leq (1-\delta)|\mathrm{Re}\,a_{ii}|, \quad i = 1,2,\cdots,n_+ + n_- \tag{6b}$$

for some $0 < \delta < 1$,

$$|\mathrm{Im}a_{ii}| \le \rho|\mathrm{Re}a_{ii}| \; i = 1,2, \cdots, n_+ + n_- \tag{6c}$$

for some constant $\rho = O(1)$, and

$$||A_{11}^{-1}A_{12}|| \le K_o, \quad ||A_{2j}|| \le K_o, \; j = 1,2, \tag{6d}$$

where $K_o > 1$ is a constant with $hK_o << 1$. If $A(x)$ is essentially diagonally dominated, then the problem (5) is said to be in *essentially diagonally dominant* (EDD) *form*.

Remarks: Assumptions (6a) and (6d) essentially say that there are n_+ equations with large positive coefficients, n_- equations with large negative coefficients, and n_o equations with coefficients which are $O(1)$. Thus the first $n_+ + n_-$ equations can be thought of as the singularly perturbed ones; the rest are not. Assumption (6c) means that the stiff behavior of the problem will be dominated by exponentially decaying or growing behavior, as opposed to highly oscillatory behavior. (If the problem is truly highly oscillatory, then other methods than the ones discussed here would be more appropriate for their treatment; see e.g. Scheid [8].) Assumption (6b) is important because it means that the equations with large coefficients behave essentially like scalar equations; this is used in the derivation of a priori estimates for the problem; see [7].

The importance of these assumptions is their implication of the following

THEOREM 1: If (5) is in essentially diagonally dominant form and there are constants $K_1 > 1$ and $K_2 > 1$ with $hK_1 >> 1$, $hK_2 >> 1$ such that

$$\begin{aligned} &||A_{11}^{-1}d^\nu A_{1j}/dx^\nu|| \le K_1, \\ &||d^\nu A_{2j}/dx^\nu|| \le K_1, \\ &||A_{11}^{-1}d^\nu f/dx^\nu|| \le K_1 \\ &||d^\nu g/dx^\nu|| \le K_1, \\ &j = 1,2, \;\; \nu = 1,2, \cdots, p \end{aligned} \tag{7}$$

and

$$\begin{aligned} &|a_{ii}(0)| \le K_2, \; i = 1,2, \cdots, n_- \\ &|a_{ii}(1)| \le K_2, \; i = n_-+1, n_-+2, \cdots, n_-+n_+, \end{aligned} \tag{8}$$

then there is a constant $C = C(K_o, K_1, K_2, \rho, \delta)$ such that the derivatives of the solution can be estimated by

$$||\frac{d^\nu y}{dx^\nu}|| + ||\frac{d^\nu w}{dx^\nu}|| \le C\Big(||y|| + ||w|| + 1\Big), \;\; \nu = 1,2, \cdots, p. \tag{9}$$

Proof: See [7].

The usefulness of this theorem is that it tells us how to construct a computational mesh which will resolve any potential unsmooth behavior in the solution to a problem in EDD form. As an example, consider the scalar problem

$$dy/dx = -a(x)y(x)\,, \quad \mathrm{Re}\,a(x) > 0, \quad 0 \le x, \qquad (10)$$
$$y(0) = y_o.$$

We construct the computational mesh for this problem by looking locally near each $x = x_o$ for a new independent variable $\tilde{x}$ defined by $x = x_o + \varphi\tilde{x}$. With this change of variables, (10) becomes

$$dy/d\tilde{x} = -\varphi a(x_o + \varphi\tilde{x})y. \qquad (11)$$

For (11), the conditions (7) reduce to

$$\left|\left|\frac{\varphi\cdot da(x_o+\varphi\tilde{x})/d\tilde{x}}{1 + \varphi|a(x_o+\varphi\tilde{x})|}\right|\right| \le K_1 \qquad (12)$$

for $p = 1$. Given a value of the threshold constant K_1, it is clear that φ can easily be determined numerically for each x_o, since $a(x)$ is a known function. We do this near a set of discrete points away from the boundaries of I, and use a uniform mesh with respect to the variable $\tilde{x}$. The details are discussed in [7].

The possible boundary layers are resolved by choosing φ at $x = 0,1$ such that conditions (8) are enforced. An exponential increase in meshsize as we move away from the boundaries is then allowed by conditions (7). Again, the implementation of this is discussed in more detail in [7].

3. Transformation to essentially diagonally dominant form. Although the results of section 2 have been stated only for problems that are in EDD form, they can easily be generalized to include problems which can be put into this form by a smooth transformation. We have

THEOREM 2: If a transformation matrix function $S(x)$ with

$$||d^\nu S(x)/dx^\nu|| \le K_1,\ \nu = 1,2,\ldots,p \qquad (13)$$

exists such that $A(x) := S^{-1}(x)\tilde{A}(x)S(x)$ is essentially diagonally dominated, and conditions (7) and (8) hold with f, g defined by $(f,g)^T := S^{-1}\tilde{f}$, then the derivatives of the solutions of

$$\frac{dv(x)}{dx} = \tilde{A}(x)v(x) + \tilde{f}(x). \qquad (14)$$

can be estimated by

$$\left|\left|\frac{d^\nu v}{dx^\nu}\right|\right| \le C\Big(||v|| + 1\Big),\ \nu = 1,2,\cdots,p. \qquad (15)$$

If for every x_o, the eigenvalues κ_i of $A(x_o)$ can be grouped into well separated sets, then it is possible to construct such a transformation locally, and this can be done automatically as part of a computer code. As with conditions (7), condition (13) just becomes a restriction on the computational mesh. The details of the construction of such a transformation can be found in [7].

The method we use there involves the solution of algebraic matrix Ricatti equations by iterative techniques.

By the definition above of EDD form, n_+ is the number of large positive eigenvalues, n_- is the number of large negative eigenvalues, and n_o is the number of moderately-sized eigenvalues of A. Here "large" and "moderately-sized" mean with respect to the local meshsize h. There is no reason to expect that the number of eigenvalues in each of these groups will be constant as a function of the location x in I. In fact, it is appropriate to define a *turning point region* as a subinterval of I in which one or more eigenvalues change their orders of magnitude. If we assume for the moment that near a given point x_o, the three sets of eigenvalues $\{\kappa_i^+\}$, $\{\kappa_i^-\}$ and $\{\kappa_i^o\}$ satisfy

$$\frac{1}{n_+}\sum \kappa_i^+ = +1\cdot O(\frac{1}{\epsilon}), \quad \frac{1}{n_-}\sum \kappa_i^- = -1\cdot O(\frac{1}{\epsilon}), \quad \frac{1}{n_o}\sum \kappa_i^o = O(1),$$

where $0 < \epsilon << 1$, then if any of the eigenvalues change their order of magnitude, they must do so in such a way that the smoothness conditions (7) are violated. Thus we expect such a region to be one in which the solution can change quickly.

By establishing arbitrary threshold constants, we can for every x_o in I group the eigenvalues of $A(x_o)$ into one of these three groups. The grouping will change when eigenvalues change their order of magnitude with respect to the local meshsize h. We can expect that the interval I will be divided up into several subintervals, on each of which the eigenvalues stay in their respective groups. On each such subinterval the system of equations (14) will be transformed to a corresponding block structure. In our code, this procedure is carried out automatically.

4. Difference approximations. In the examples we have computed, we have used a combination of one-sided and centered two-point schemes. Once the system of equations has been transformed to EDD form (5),(6), we use the implicit Euler's method on the first n_- equations, the explicit Euler's method on the next n_+ equations, and the Trapezoidal Rule on the remaining n_o equations. An interesting result is that although we have used one-sided schemes on some of the equations, this approximation gives second order accurate solutions:

THEOREM 3: Let (5) be approximated as described above, assume that conditions (6), (7) and (8) hold, and that $h||A_{11}|| >> 1$. Then denoting by u the approximation to y, and z the approximation to w, we have the error estimate

$$\begin{aligned}(||y-u|| + ||w-z||) \\ \leq const.\Big(h^2 + |y^-(0) - u^-(0)| + |y^+(1) - u^+(1)| + |w(0) - z(0)|\Big).\end{aligned} \tag{16}$$

Here the partition (y^-,y^+) of the variables corresponds to the equations with negative and positive large diagonal elements, respectively.

Proof: Again, see [7].

5. Nonlinear problems with internal layers. We turn now to nonlinear problems. Recently we have been studying the behavior of our numerical method when applied to singularly perturbed second order ODEs that have solutions with internal layers. Our procedure for solving linear problems, discussed above, can be thought of as a technique for *resolving* the solutions of stiff linear ODEs, i.e. the numerical solution we get is everywhere a good approximation to the analytic solution. An approach for solving nonlinear ODEs is to embed this method within a Newton iteration procedure for solving the nonlinear (continuous) ODE of interest. Let u, a, $\gamma \in R^n$, $B_j \in R^n \times R^n$, $j = 0,1$, and consider the system of nonlinear first order ODEs given by

$$u_x(x) = a(u(x)), \tag{17a}$$

where for simplicity we specify linear boundary conditions

$$B_0 u(0) + B_1 u(1) = \gamma. \tag{17b}$$

Here subscripts denote (partial) differentiation. Let $u^n(x)$ be a previous guess or approximate solution to (17), and linearize (17) about this function. We obtain

$$L(u^n(x),\epsilon)\hat{u}(x) := \hat{u}_x(x) - a_u(u^n(x))\hat{u}(x) = -a(u^n(x)) - u_x^n(x). \tag{18}$$

Here $\hat{u}(x) := u^{n+1}(x) - u^n(x)$ is the correction to the guess u^n. The problem (18), (17b) is clearly a linear one of the form we discussed above, and it can therefore our method can be applied to it. (For each linearized problem, a transformation to EDD form and a stretching function is determined, and the difference approximation is applied.) The Newton procedure is then, as usual, to start with some initial guess $u^0(x)$ and then solve (18),(17b) for $n = 0,1,2,...$ until convergence, measured in some appropriate way, is obtained. Given a sufficiently good initial guess, one expects this procedure to converge rapidly.

Unfortunately, the Newton iteration procedure which has been briefly outlined above is not necessarily a good one. For some singularly perturbed problems with internal layers, we have found that this procedure often fails to converge, even when an initial guess very close to a known isolated solution is used. Although we believe that the convergence properties of the standard Newton procedure could be improved by using continuation with sufficiently small steps, or by the use of some kind of damping procedure, an analysis of the underlying mathematical difficulties with layer problems has lead us to another technique. The essential reason for the failure of the usual Newton iteration to converge is that if u is a solution of (17) with an internal layer, the spectrum of the linearized operator $L(u(x),\epsilon)$ contains one small (e.g. $O(\epsilon)$) eigenvalue which is well separated from the rest of the spectrum. Thus the linearized operator is nearly singular, and difficulties with Newton can be expected. It should be emphasized that this difficulty lies with the continuous problem, and is not an artifact of the numerical method that is used.

In order to understand this difficulty, we restrict our consideration for the moment to 2nd order singularly perturbed ODEs of the form

$$\epsilon^2 u_{xx} + b(x,u) = 0, \quad 0 \leq x \leq 1, \tag{19}$$

where for boundary conditions we specify $u(0)$ and $u(1)$. Assume for the moment that problem (19) has a solution with an internal layer of width $O(\epsilon)$, and no boundary layers. More specifically, we assume that a function $\bar{u}(x)$ exists which satisfies (19) and has the properties that

$$|d\bar{u}/dx| \leq \begin{cases} c_o & \text{for } |x - x_o| \geq c_1\epsilon \\ \dfrac{c_2}{\epsilon} & \text{for } |x - x_o| \leq c_1\epsilon. \end{cases} \tag{20}$$

Here c_j, $j = o,1,2$ are $O(1)$ constants. The "internal layer ", if it exists, is located near $x = x_o$. Introducing a stretched independent variable t defined by $x = x_o + \epsilon t$, (19) becomes

$$u_{tt} + b(x_o + \epsilon t, u) = 0. \tag{21}$$

Now linearize (21) about $\bar{u}$. Letting $u := \bar{u} + w$, we obtain

$$L(\bar{u},\epsilon)w := w_{tt} + b_u(x_o + \epsilon t, \bar{u})w = 0 \tag{22}$$

Since (22) is "nearly" an autonomous system, the derivative of $\bar{u}$ will very nearly satisfy (22). To verify this, let $v := \bar{u}_t$; then by differentiating (21) we obtain

$$L(\bar{u},\epsilon)v = \; = -\epsilon b_x(x_o + \epsilon t, \bar{u})). \tag{23}$$

By the characterization (20) of $\bar{u}$, we see that v satisfies (23) with boundary conditions

$$v(0) = O(\epsilon), \quad v(1/\epsilon) = O(\epsilon), \tag{24}$$

and so is apparently "very nearly" an eigenfunction of the linearized operator $L(\bar{u},\epsilon)$ with eigenvalue 0. We interpret this as meaning that the operator $L(\bar{u},\epsilon)$ has at least one very small eigenvalue.

This argument is certainly only a heuristic one at best, but it makes the existence of a small eigenvalue quite plausible. It also shows that the existence of the small eigenvalue is connected with the occurence of an internal layer in the solution of the original problem. A more rigorous presentation will appear in [3]. We also note that a similar argument can be made for equations of the form

$$\epsilon u_{xx} + f(u)_x + b(x,u) = 0.$$

Because of this special structure of linearized problem, a straight-forward generalization of Newton's method is possible. In order to discuss this, we consider the solution by Newton's method of the system of ordinary differential equations

$$F(x,u(x),u_x(x)) = 0, \quad F,u \in R^n.$$

Linearization gives the iteration procedure

$$\begin{aligned} &F(x,u^k,u_x^k) + F_u(x,u^k,u_x^k)(u^{k+1} - u^k) \\ &\quad + F_{u_x}(x,u^k,u_x^k)\frac{d}{dx}(u^{k+1} - u^k) = 0, \quad k = 0,1,2,... \end{aligned} \tag{25}$$

where some initial guess u^o must be specified. The discretization of (25) on a set of N meshpoints will lead to a system of equations for the iterates $y^k \in R^{Nn}$ given by

$$A(y^k)(y^{k+1} - y^k) = b(y^k) \tag{26}$$

where $A \in R^{Nn \times Nn}$. The matrices $A_k := A(y^k)$ can be easily and cheaply monitored for the existence of an eigenvalue of small magnitude well separated from the rest of the spectrum by taking a few steps of inverse iteration once the LU-decomposition of A has been computed. Typically, for an arbitrary initial guess y^o, the first few A_k will not have this property. This is because the first A_k do not approximate the continuous linearized operator well. As the solution begins to approximate the internal layer well, we will arrive at an iterate A_k with such a small eigenvalue, and the inverse iteration will converge very quickly; we also obtain as a byproduct the associated left and right eigenvectors ψ_k^T and φ_k. (The superscript T denotes the transpose of a vector.) If for each k, the vectors $b(y^k)$ satisfy

$$\psi_k^T \cdot b(y^k) = 0 \tag{27}$$

at least approximately, then there will be no difficulty in continuing with the Newton iteration procedure. This often happens in practice; an example of such a problem is given below (figure 1). However, we cannot always expect to be so lucky. If (27) does not hold for the problem of interest, then we replace the iteration (26) with the following: Let $P_k := \varphi_k \psi_k^T$, and compute the iterates y_{m+1}^k by the iteration

$$\begin{aligned} &(I - P_k)A(y^k)(y_{m+1}^k - y_m^k) = (1 - P_k)b(y_m^k), \quad m = 0,1,2,... \\ &y_o^k := y^k. \end{aligned} \tag{28}$$

By the definition of the iteration procedure (28), each iterate y_m^k has the same component in the direction of the eigenvector φ_k. In fact this is the motivation for choosing this iteration: since there is no component of the right-hand side of (28) in the direction of the eigenvector, the iteration can be expected to converge with no problem. However, the component of y in this direction still needs to be chosen somehow. It is clear that the remaining free parameter in the problem must be chosen so that the solution y satisfies

$$\psi_k^T \cdot b(y) = 0.$$

Denote by y_∞^k the final iterate of (28). The procedure we use is to try to choose a correction factor $\sigma_k := \sigma_k(y_\infty^k, y^k)$ such that $y^{k+1} := y_\infty^k + \sigma_k \varphi_k$ will satisfy

$$|\psi_{k+1}^T f(y_\infty^k + \sigma_k \varphi_k)| \leq |\psi_k^T f(y^k)|. \tag{29}$$

We do this by using a slight modification of "regula falsi" on the nonlinear equation

$$\psi^T(y) \cdot b(y) = 0,$$

where $\psi^T(y)$ is the appropriate left eigenvector of $A(y)$.

6. Examples. In this section we briefly present some numerical computations. More examples can be found in [7] and will be included in [3].

As a first example we show a computation of one of the (multiple) solutions of the equations for one-dimensional steady gas flow in a Laval nozzle. (All variables but the velocity have been eliminated, and the equations have then been "regularized" to give a singular perturbation problem.) The equations, boundary conditions and a plot of a computed solution (physically unstable, incidentally) is shown in figure 1. The main point of this computation was to verify that our procedure for automatically decoupling the scales and stretching the independent variable worked properly for nonlinear singular perturbation problems. Thus, as an initial guess, an asymptotic solution for a large value of ϵ was used, and continuation in ϵ was used to compute solutions for smaller values of ϵ. In order to illustrate the division of the interval into subintervals with different block structure (c.f. section 3), a solution for an intermediate value of ϵ is illustrated in figure 1. The automatically constructed mesh is indicated by '+' marks on the horizontal lines in the plot. The vertical lines in the interior of the interval indicate the endpoints of subintervals with different block structure. Note that although this (nonlinear) problem has an internal layer, there is no difficulty associated with using the usual Newton procedure because condition (27) turns out to hold.

Figure 2 shows the result of the numerical computation by the iteration technique discussed in section 5 of a problem where (27) is violated. The problem is given by

$$\begin{aligned} &\epsilon^2 u'' - u(u-1)(u-\theta(x)) = 0, \quad 0 \leq x \leq 1, \\ &u(0) = 1/4, \quad u(1) = 2, \quad \theta(x) = 5/2 - x. \end{aligned} \tag{30}$$

In the nonlinear iteration procedure, an asymptotic expansion for a known solution with an internal layer was used as an initial guess. Usual Newton was used for the first three iterations, at which point the difficulty associated with a small eigenvalue of the linearized problem arose. The modified iteration procedure discussed above was then used to get a converged solution.

More details will appear in [3].

Acknowledgements. I am grateful to Heinz Kreiss for the many illuminating discussions we have had on this problem. I also acknowledge the National Science Foundation for its generous support of this work under contract no. DMS-8312264.

REFERENCES

[1] ASCHER, U., On some difference schemes for singular singularly-perturbed boundary value problems, University of British Columbia Dept. of Comp. Sci. Rept. (1983).

[2] ASCHER U. and R. WEISS , Collocation for singular perturbation problems I: First order systems with constant coefficients, SIAM J. Num. Anal., 20, pp. 537-557, (1983).

[3] BROWN, D.L., W.L. KATH and H.O. KREISS, to appear.

[4] KEVORKIAN, J. and J. COLE, Perturbation methods in applied mathematics, Springer-Verlag (1981).

[5] KREISS, H.O., Centered difference approximations for singular systems of ordinary differential equations, Istituto Nazionale di Alta Matematica, Symposia Matematica, Volume X, pp. 454-465, (1972).

[6] KREISS, H.O., Central difference schemes and stiff boundary value problems, m.s. (1984).

[7] KREISS, H.O., N. NICHOLS and D. L. BROWN, Numerical methods for stiff two-point boundary value problems, University of Wisconsin, MRC Rept. #2599, November 1983, to appear in SIAM J. Num. Anal.

[8] SCHEID, R. E., The accurate numerical solution of highly oscillatory ordinary differential equations, Math. of Comp., 41, pp. 487-509, (1983).

[9] WEISS, R., An analysis of the box and trapezoidal schemes for linear singularly perturbed boundary value problems, Math. of Comp., 42, pp. 41-67, (1984).

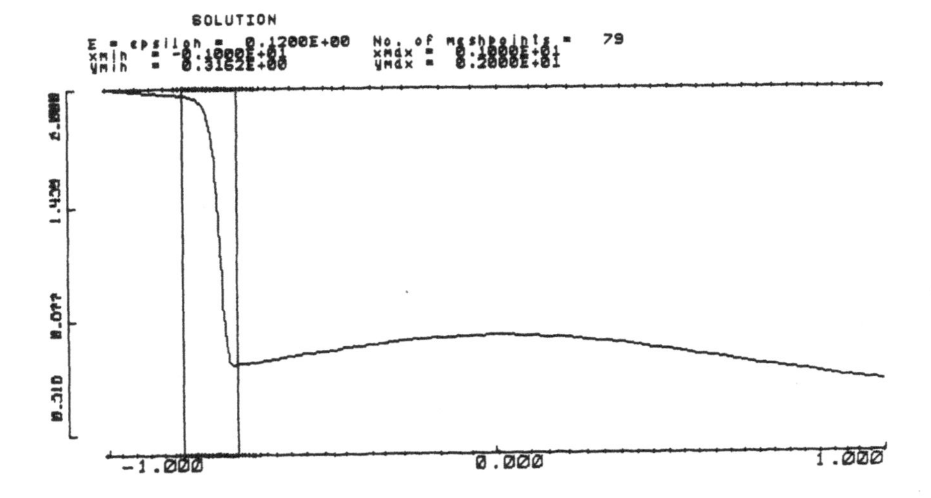

Fig. 1: A solution of

$$\varepsilon u'' + \left(\frac{\gamma+1}{\gamma-1} - \frac{2B}{u^2}\right)u' - \frac{A'(x)}{A(x)}\left(\frac{2B}{u} - u\right) = 0, \quad -1 \;\; x \;\; 1,$$

$$u(-1) = 2, \quad u(1) = .58\ , \quad B = 3.86, \quad A = 1+\frac{1}{5}\cos\left(\frac{\pi}{2}(x+2)\right)$$

Vertical lines indicate endpoints of subintervals of different matrix block structure.

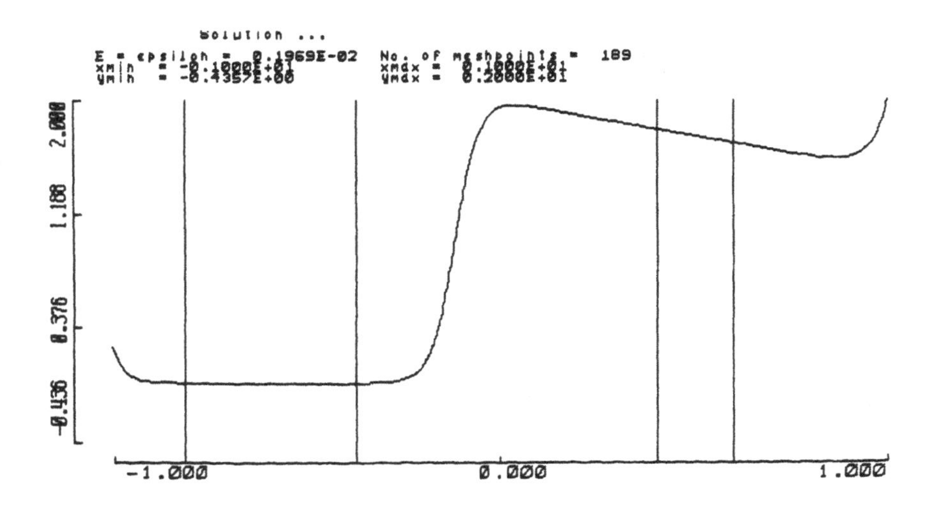

Fig. 2: A solution of (30) with u(-1) = 1/4, u(1) = 2.

Progress in Scientific Computing, Vol. 5
Numerical Boundary Value ODEs

Numerical solution of singular perturbed boundary value problems using a collocation method with tension splines

Maximilian R. Maier

1. Introduction

A collocation method is given for the numerical solution of a singular perturbed boundary value problem of the form:

$$
\begin{aligned}
&y' = f(x,y,\varepsilon) && a \le x \le b \\
&r(y(a),y(b)) = 0 && \\
(1.1)\qquad &y=(y_1,\dots,y_n)^t && y_i: [a,b] \to R \\
&f=(f_1,\dots,f_n)^t && f_i: [a,b] \to R \\
&r=(r_1,\dots,r_n)^t && r_i: R \times R \to R
\end{aligned}
$$

Here ε is a small positive parameter and the solutions of (1.1) generally behave nonuniformly as ε tends towards zero. In the following the small parameter ε is skipped, because the method does not require any explicit information about it. The solution of (1.1) generally has a multiscale character, i.e. there are thin regions, called boundary layers, where it varies rapidly, while away from the boundary layers it behaves regularly. There is a large literature on the theoretical study of singularly perturbed boundary value problems of which only O'Malley[16] and Howes[12] are mentioned. Such problems have also been studied numerically and there are methods based on finite differences (cf. Abrahamsson, Keller, Kreiss[2], Kreiss[13]

Pearson[17,18]), and methods based on singular perturbation theory (cf. Miranker[15], Flaherty, O'Malley[11]). A collocation method using both cubic polynomials and tension splines for second-order linear singularly perturbed boundary value problems is described in Flaherty, Mathon[10]. The collocation method described here is based on an algorithm of Dickmanns, Well[9], where the solution of (1.1) is approximated by piecewise cubic Hermite polynomial functions. This allows a local construction of the polynomials in every subinterval, which makes the method easy to program. Here the cubic polynomials are replaced by tension splines inside the layers. The tension parameters are selected from the eigenvalues of the functional matrix $D_y f$ of (1.1). Inside the boundary layers the solution is determined by eigenvalues with absolutely large real parts and therefore one of these eigenvalues is chosen as tension parameter. In the outer solution region zero is taken for the tension parameter, which leads to cubic polynomials in this region. This strategy is described in chapter three. In chapter two the collocation method is constructed together with a stepsize control, which proved to give reliable results in all tested examples. In chapter four the advantages of using tension splines instead of cubic polynomials is demonstrated on a simple example. Besides this three problems arising in the physical theory of semiconducting devices are solved.

2. Collocation method

The method uses Lobatto points for collocation points, which gives good stability for singularly perturbed differential equations, see Ascher, Weiss[4] and makes it easy to determine the exponential spline in every subinterval. One first introduces a partition of the interval [a,b]:

$$a=x_1 < x_2 < \dots < x_{m-1} < x_m = b \tag{2.1}$$

In every subinterval $[x_i,x_{i+1}]$ $i=1,..,m-1$ the solution of

(1.1) is approximated by a tension spline e(x) of the form, see Rentrop[19]:

$$e(x) = z_{i+1}t + z_i(1-t) + \frac{d_{i+1}}{\lambda_i^2}\left(\frac{\sinh \mu_i t}{\sinh \mu_i} - t\right) + \frac{d_i}{\lambda_i^2}\left(\frac{\sinh \mu_i (1-t)}{\sinh \mu_i} - (1-t)\right)$$

(2.2)

with $t = \frac{x-x_i}{h_i}$ $\quad h_i=x_{i+1}-x_i$ $\quad \lambda_i$ tension parameter

$\mu_i = \lambda_i h_i$ $\quad z_{i+1}, z_i$ approximations for $y(x_{i+1}), y(x_i)$

(2.2) allows the computation of e(x) in a stable and efficient manner, including the limiting cases $\lambda_i \to 0$, $\lambda_i \to \infty$. d_{i+1} and d_i are determined by the conditions:

$$(2.3) \qquad e'(x_i)=z'_i \qquad e'(x_{i+1})=z'_{i+1} \qquad z'_i=f(x_i,z_i)$$

The resulting spline function is continuous in the first derivative and satisfies the differential equations at the nodes x_i. For the collocation equations e(x) and e'(x) are computed at the centre $\bar{x}_i:=0.5(x_i+x_{i+1})$ of each subinterval:

$$(2.4) \qquad F_i := F_i(z_i,z_{i+1}) = E'_i - f(\bar{x}_i,E_i) = 0 \qquad i=1,..,m-1$$

$$E_i := e(\bar{x}_i), \qquad E'_i := e'(\bar{x}_i)$$

Together with the boundary condition:

$$(2.5) \qquad F_m := F_m(z_1,z_m) = r(z_1,z_m) = 0$$

one gets the generally nonlinear system

$$(2.6) \qquad F(z) = 0 \qquad F=(F_1,F_2,..,F_m)^t \qquad z=(z_1,z_2,..,z_m)^t$$

which is solved iteratively by a modified Newton method (cf. Stoer, Bulirsch[25]):

$$(2.7) \qquad z^{(k+1)} = z^{(k)} - \tau_k \cdot DF(z^{(k)})^{-1} F(z^{(k)}) \qquad k=1,2,..$$

This leads to a linear system, where the matrix has the block form:

$$(2.8) \qquad DF(z) = \begin{bmatrix} G_1 & \bar{G}_2 & & & & \\ & G_2 & \bar{G}_3 & & 0 & \\ & & \ddots & \ddots & & \\ & 0 & & \ddots & \ddots & \\ & & & & G_{m-1} & \bar{G}_m \\ A & & & & & B \end{bmatrix}$$

$$G_i = \frac{\partial F_i(z_i, z_{i+1})}{\partial z_i} \qquad \bar{G}_{i+1} = \frac{\partial F_i(z_i, z_{i+1})}{\partial z_{i+1}}$$

$$A = \frac{\partial r(z_1, z_m)}{\partial z_1} \qquad B = \frac{\partial r(z_1, z_m)}{\partial z_m}$$

The resulting linear equation system is solved by Householder transformations with respect to the sparse structure of (2.8). G_i, $\bar{G}_{i+1}$ are computed by numerical differentiation. In certain iteration steps they are approximated by rank-1 approximations (see Broyden[6]), which reduces the amount of work. This strategy together with the estimation of the relaxation factor τ_k in (2.7) is described in Deuflhard[7,8].

In order to obtain a reliable method a stepsize control is necessary. In Russell, Christiansen[22] various possibilities are described. In the present case the stepsize control is based on the following theorem, which can be proved by repeatedly applying Rolle's Theorem.

Theorem (2.9)

Suppose that $g\in C^{(4)}[a,b]$, $e(x)$ tension spline with tension parameter λ and $e(a)=g(a)$, $e'(a)=g'(a)$, $e(b)=g(b)$, $e'(b)=g'(b)$. Then there exists for every $x\in[a,b]$ a $\xi\in(a,b)$ with:

$$g(x) - e(x) = \frac{(x-a)^2(x-b)^2}{4!} \left(g^{(4)}(\xi) - \lambda^2 e''(\xi) \right)$$

This leads to the following suggestion for the stepsize $h_i=x_{i+1}-x_i$:

$$(2.10) \qquad h_i = \sqrt[4]{\frac{384\cdot TOL}{EST}}$$

TOL: prescribed tolerance

EST: error estimate $EST = ||g^{(4)}(\xi)-\lambda_i^2 e''(\xi)||_\infty$

The value of EST is computed by setting:

$\xi=\bar{x}_i=\frac{1}{2}(x_i+x_{i+1})$

λ_i: actual tension parameter

$e''(\xi)$ from (2.2)

$g^{(4)}(\xi) = \lambda_i e'''(\xi)$

In the case $\lambda=0$ $g^{(4)}(\xi)$ is estimated as in de Boor[5]. New nodes are inserted only in certain iteration steps. Starting with the initial mesh (2.1), the Newton method iterates until convergence. Then in every subinterval new tension parameters and estimated stepsizes h_i are determined and, if necessary, new nodes are inserted. Then the Newton iteration starts again with the new mesh, leading to an improved solution. This process is repeated until no further nodes must be inserted.

3. Selection of the tension parameter

The tension parameter λ_i in the interval $[x_i,x_{i+1}]$ $i=1,..,m$ is selected from the eigenvalues EV of the functional matrix $D_y f$ of (1.1) inside the boundary layers, while in the outer solution region the algorithm takes $\lambda_i=0$ which saves the computations of the eigenvalues. For a singularly perturbed

boundary value problem there are some eigenvalues with absolutely very large realparts and some of moderate size. the eigenvalues with the large realparts cause the rapid variation of the solution inside the boundary layers, and are therefore selected as tension parameter. A further restriction for the selection of the tension parameter is, that inside the left boundary layer only eigenvalues with negative and inside the right boundary layer only eigenvalues with positive realparts have to be considered.
In a first step an estimate Λ_i for the eigenvalue is computed from the values $z_i=(z_{i1},z_{i2},\ldots,z_{in})^t$ at the nodes x_i, i=1,..,m. Assuming the solution of (1.1) locally behaves like the exponential function one gets:

$$(3.1) \qquad \Lambda_i = \max_j \left| \ln \left| \frac{z_{i+1,j}}{z_{i,j}} \right| \right| / h_i \qquad h_i=x_{i+1}-x_i$$

This value indicates, if zero or one of the eigenvalues with large realparts should be chosen as tension parameter. In the case that one of the eigenvalues with large realpart is selected, the eigenvalue with the smallest relative deviation from (3.1) is taken:

$$(3.2) \qquad \lambda_i = \lambda : \frac{||\lambda| - \Lambda_i|}{\max(|\lambda|,\Lambda_i)} \qquad \text{minimal for } \lambda \in EV$$

As mentioned before, a further restriction for the decision in (3.2) is that inside the left boundary layer only λ with negative and in the right boundary layer only λ with positive realparts have to be considered.

4. Numerical results

The following computations were performed in FORTRAN IV with single precision (48 bit mantissa) on the CDC CYBER 175 of the Leibniz Rechenzentrum der Bayerischen Akademie der Wissenschaften. The abbreviations TOL, CT, NFC, M denote respectively required relative accuracy, computer time in seconds, number of function calls of the right side of (1.1), and number of nodes of the final mesh.

Example 1

(4.1.1) $\qquad \varepsilon u'' + u' + u^2 = 0 \qquad u(0)=0 \quad u(1)=0.5$

The solution of this problem can be approximated by

(4.1.2) $\qquad u(x,\varepsilon) \sim \frac{1}{x+1} - \exp(-x/\varepsilon) + O(\varepsilon)$

There is a boundary layer of width $O(\varepsilon)$ near $x=0$. On this example it is demonstrated how much better the tension spline behaves on singularly perturbed boundary value problems than a cubic spline. Fig. (4.1.3) shows for $\varepsilon=10^{-2}$ the tension spline with tension parameter from chapter 3 (dashed line) and the cubic polynomial (straight line) for a uniform mesh of 11 nodes. The oscillations continue far into the outer solution region. They can be damped by using more nodes inside the boundary layer. In Fig.(4.1.4) the mesh is refined by placing five nodes equidistant inside the interval [0,0.2]. The oscillations of the cubic polynomials are damped, but there is still oscillation in the region where the boundary layer meets the outer solution.

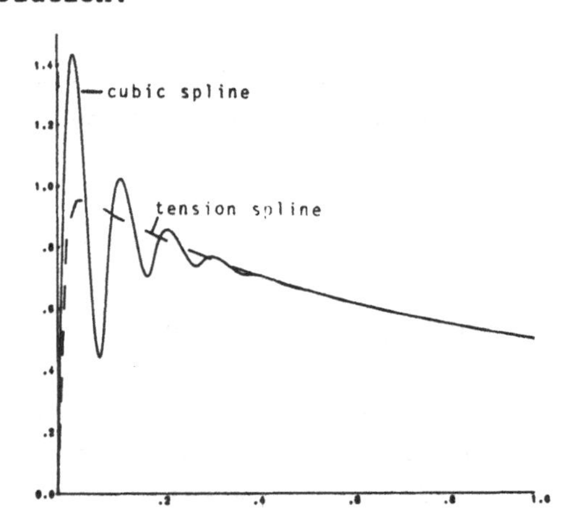

Fig.(4.1.3):

x_i: 0 , 0.1, 0.2, 0.3, ... , 0.9, 1.

λ_i: 100, 0 , 0 , 0 , ... , 0

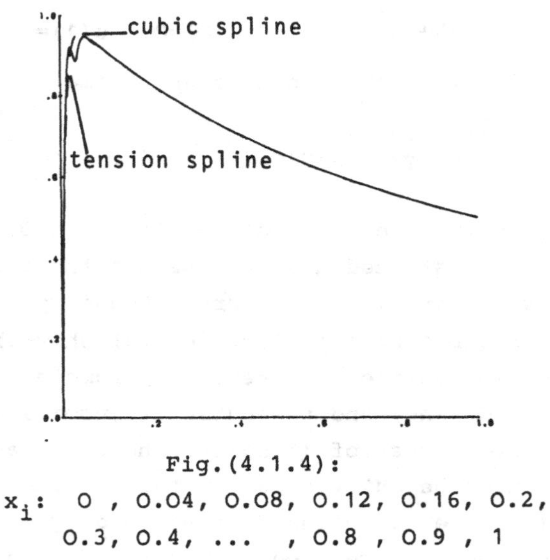

Fig.(4.1.4):

x_i: 0 , 0.04, 0.08, 0.12, 0.16, 0.2, 0.3, 0.4, ... , 0.8 , 0.9 , 1

λ_i: 100, 0 , 0 , 0 , . . .

Example 2

This singularly perturbed system from Vasil'eva, Stel'makh [26] describes a simplified model of a symmetric semiconductor diode:

$$(4.2.1) \qquad \begin{aligned} \varepsilon E' &= p - n + N \\ p' &= pE - J \\ n' &= -nE + J \end{aligned} \qquad \begin{aligned} &p(0)=n(0) \\ &p(1)=0 \\ &n(1)=N \end{aligned}$$

with E electric field strength, p,n hole and electron density, N doping profile, J current density N=J=1), ε small parameter proportional to the ratio of the space charge region to the length of the diode.
An asymptotic analysis of (4.2.1) is done in [26] and for more general boundary conditions in Smith[24]. The solution has a boundary layer of width $O(\sqrt{\varepsilon})$ at x=0. It is shown for $\varepsilon=10^{-3}$, $\varepsilon=10^{-4}$ in Fig.(4.2.3) and Fig.(4.2.4). The computations with the collocation method were done with a pres-

cribed tolerance TOL=10^{-4} for $\varepsilon=10^{-3}$, 10^{-4},..., 10^{-12}. The amount of computation required is shown in (4.2.2):

ε	10^{-3}	10^{-4}	10^{-5}	10^{-6}	10^{-7}	10^{-8}	10^{-9}	10^{-10}	10^{-11}	10^{-12}
CT	0.36	0.5	0.65	0.57	0.57	0.67	0.7	0.72	0.72	0.83
NFC	1411	1997	2735	2330	2435	3019	3245	3245	3245	3861
M	32	32	42	56	62	65	66	66	66	66

Table (4.2.2)

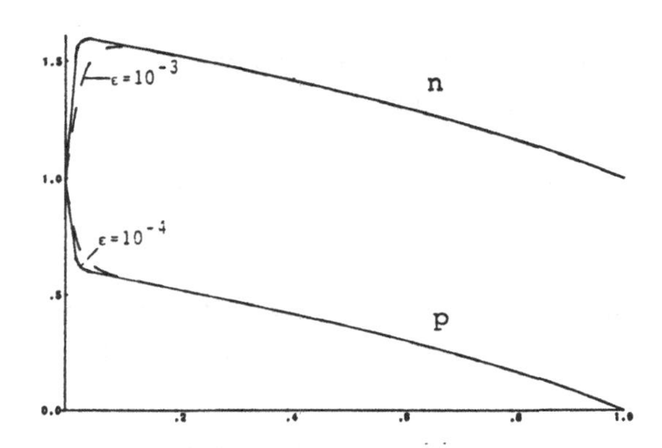

Fig.(4.2.3): Hole and electron densities p,n

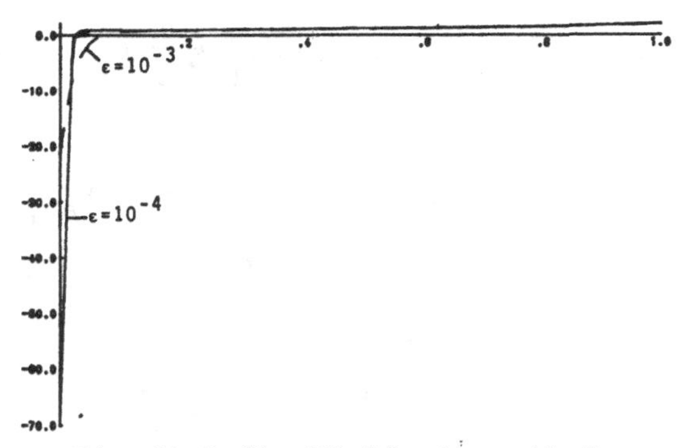

Fig.(4.2.4): Field strength E

For small ε the number of nodes is almost constant, whereas by using cubic splines for $\varepsilon<10^{-5}$ more than 100 nodes are needed, which substantially increases the amount of required computer time. Using the same mesh with M from (4.2.2) for cubic splines yields solutions which have no correct digit in the region where the boundary layer meets

the outer solution.
The computed results were compared with the results from Ascher[3] and Maier, Smith[14], which showed an accuracy of 4-7 decimals.

Example 3

The boundary value problem (4.3.1) simulates the electrical properties of electron-irradiated silicon. It is described in Sigfridsson, Lindström[23] and solved by a difference method in Abrahamsson[1]. One has the problem:

$$(4.3.1) \qquad \begin{aligned} I\cdot y' &= F(x,y,z) \qquad & y(0)&=1 \\ I\cdot z' &= G(x,y,z) \qquad & z(1)&=0 \end{aligned}$$

y,z,I normalized electron, hole and current densities. The right sides of the differential equations are given as:

$$(4.3.2) \qquad \begin{aligned} F(x,y,z) &= (y+\beta z)(A_1 y\tilde{f} - \sum_{i=1}^{N_A} \hat{f}_i - \sum_{j=1}^{N_D} \bar{f}_j) \\ G(x,y,z) &= (y+\beta z)(A_1 z\tilde{f} + \frac{1}{\beta} \sum_{i=1}^{N_A} \hat{f}_i + \frac{1}{\beta} \sum_{j=1}^{N_D} \bar{f}_j) \end{aligned}$$

$$\tilde{f} = 1-y+z - \sum_{i=1}^{N_A} \gamma_{a_i}(x) \frac{y+\alpha_{a_i} K_{ap_i}}{y+K_{an_i}+\alpha_{a_i}(K_{ap_i}+z)} + \sum_{j=1}^{N_D} \gamma_{d_j}(x) \frac{K_{dn_j}+\alpha_{d_j} z}{y+K_{dn_j}+\alpha_{d_j}(K_{dp_j}+z)}$$

$$(4.3.3) \qquad \hat{f} = \alpha_{a_i} A_{an_i} \gamma_{a_i}(x) \frac{yz-K_{an_i}K_{ap_i}}{y+K_{an_i}+\alpha_{a_i}(K_{ap_i}+z)}$$

$$\bar{f} = \alpha_{d_j} A_{dn_j} \gamma_{d_j}(x) \frac{yz-K_{dn_j}K_{dp_j}}{y+K_{dn_j}+\alpha_{d_j}(K_{dp_j}+z)}$$

$$\beta = \frac{1}{3} \qquad A_1=0.05162 \qquad N_A=2 \qquad N_D=1$$

$$\alpha_{a_1}=\alpha_{a_2}=\alpha_{d_1}=1 \qquad A_{an_1}=A_{an_2}=A_{dn_1}=2.222\ 10^{-3}$$

(4.3.4) $$K_{ap_1}=1.854\ 10^{-4} \quad K_{an_1}=21.47 \qquad K_{ap_2}=0.1021$$

$$K_{an_2}=3.899\ 10^{-2} \quad K_{dp_1}=2.902\ 10^{3} \quad K_{dn_1}=1.371\ 10^{-6}$$

$$\gamma_{a_1}(x)=15 \qquad \gamma_{a_2}(x)=10 \qquad \gamma_{d_1}(x)=400 \qquad I\in[10^{-12},1]$$

The boundary value problem is interesting in two ways, first the eigenvalues are O(1/I), which yields very narrow boundary layers of width O(I), which are physically interesting, and second the solution behaves quite differently inside the boundary layers as indicated by Fig.(4.3.5).

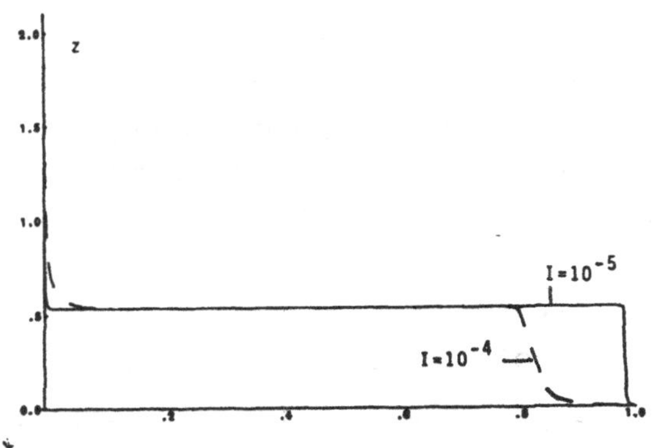

Fig.(4.3.5): normalized hole density z

At x=0 the solution is governed by an absolutely increasing negative eigenvalue, while at x=1 one eigenvalue changes its sign and causes a turning point inside the boundary layer.
For $TOL=10^{-4}$ one has the following amount of work for $I=10^{-6}-10^{-12}$:

(4.3.6) CT: 1.4-1.5 NFC: 8074-8490 M: 53-55

Example 4

A symmetric semiconductor diode (n-type zone) is described by the following boundary value problem, see Rieger[20]:

$$
(4.4.1)\qquad
\begin{aligned}
V''(x) &= -(N(x) + p(x) - n(x)) \\
\Phi_p'(x) &= -J_p(x)/(\mu_p \cdot p(x)) \\
\Phi_n'(x) &= -J_n(x)/(\mu_n \cdot n(x)) \\
J_p'(x) &= -\frac{(p(x)n(x) - 1)}{\tau_p(n(x)+1) + \tau_n(p(x)+1)} \\
J_n'(x) &= -J_p'(x)
\end{aligned}
$$

with $\quad p(x) = \exp(\Phi_p(x) - V(x)) \quad n(x) = \exp(V(x) - \Phi_n(x))$

where V is the electrostatic potential, $N(x)=N_D-N_A$ doping profile, p,n hole and electron density, J_p, J_n hole and electron current density, μ_p, μ_n carrier mobilities, τ_p, τ_n life-time parameters, Φ_p, Φ_n Quasi-Fermi-potentials. The parameter values are (scaling from [20]):

$$
(4.4.2)\qquad
\begin{aligned}
&\mu_p=\mu_n=480\ cm^2V^{-1}s^{-1}/\mu_o && \mu_o=38.61\ cm^2V^{-1}s^{-1} \\
&N_D=10^{14}\ cm^3/n_i && n_i=1.5\ 10^{10}\ cm^3 \\
&\tau_p=\tau_n=10^{-6}\ s/\tau_o && \tau_o=1.532\ 10^{-5}\ s
\end{aligned}
$$

The boundary conditions are:

$$
(4.4.3)\qquad
\begin{aligned}
V(0) &= 0 & V(L) &= \ln(N_D) - 0.5\,\frac{U_a}{u_t} \\
\Phi_n(0) &= -\Phi_p(0) & \Phi_n(L) &= V(L) - \ln(N_D) \\
J_p(0) &= J_n(0) & \Phi_p(L) &= \Phi_n(L)
\end{aligned}
$$

where $L = 50\ \mu m/L_o \qquad L_o = 3.915\ 10^{-3}$ cm length of n-zone
$u_t = 0.025875$ V $\qquad U_a$ applied voltage

System(4.4.1) does not explicitly contain a small parameter. First it is considered for the equilibrium condition:

a) $U_a = 0$

The solution for V, p, n ($\Phi_p=\Phi_n=0$, $J_p=J_n=0$) is shown in Fig.(4.4.4) and Fig.(4.4.5). The amount of computation is for $TOL=10^{-3}$: CT: 2.2, NFC: 4997, M: 32.

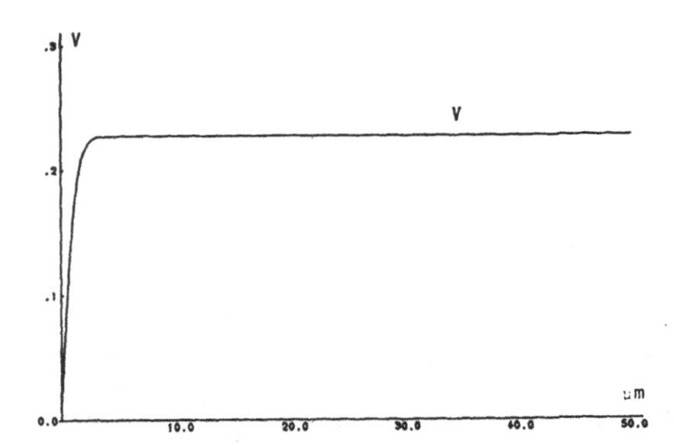

Fig.(4.4.4): Potential V for $U_a=0$

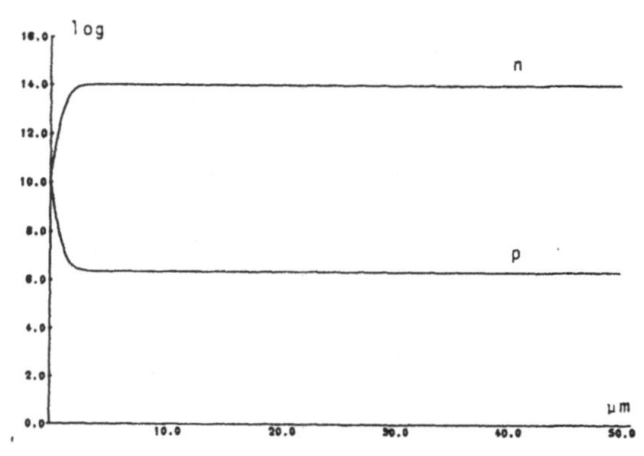

Fig.(4.4.5): Carrier densities p,n for $U_a=0$

b) Forward bias $U_a = 0.3$ V

By applying a positive voltage, the majority carriers are enabled to pass the space charge region, which results in a strong increase of the current densities J_p, J_n, see Fig.(4.4.6). The potential barrier in the middle of the diode (x=0) is reduced, which can be seen by comparing Fig.(4.4.4) with Fig.(4.4.7). Fig.(4.4.8) shows an increase of the minority carriers p in the n-zone caused by the injection from the pn-junction. The solution was determined at 64 nodes, requiring CT: 5.1, NFC: 12323.

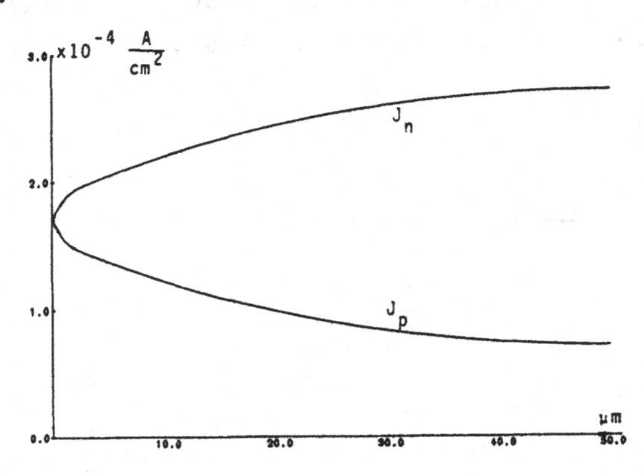

Fig. (4.4.6): Current densities J_p, J_n for U_a=0.3 V

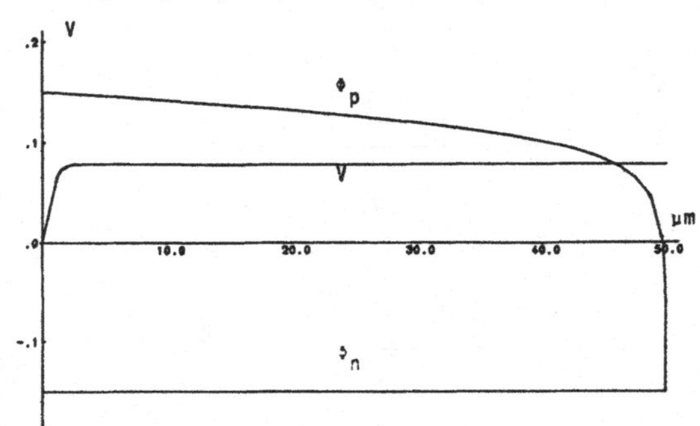

Fig. (4.4.7): Potentials V, Φ_p, Φ_n for U_a=0.3 V

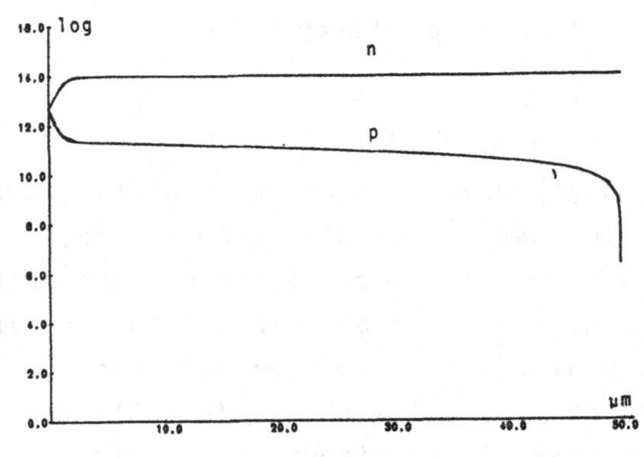

Fig. (4.4.8): Carrier densities p,n for U_a=0.3 V

c) Reverse bias $U_a = -1$ V

By applying a negative voltage the potential difference at the pn-junction at x=0 is increased. Only the minority carriers pass the space charge region, so that almost no current flows, see Fig.(4.4.9). Fig.(4.4.10) shows the increased potential difference compared to Fig.(4.4.7) and Fig.(4.4.4). The carrier densities p and n are shown in Fig.(4.4.11). Compared to Fig.(4.4.5), Fig.(4.4.8) the densities near the pn-junction at x=0 are distinctly smaller. The solution was computed by a homotopy chain, setting U_a=-0.5, U_a=-1.

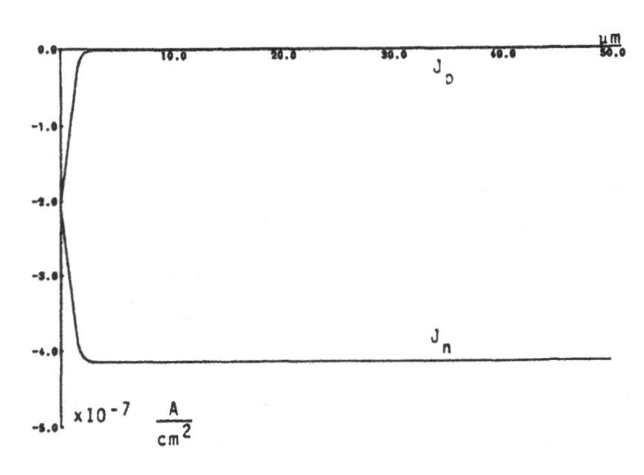

Fig.(4.4.9): Current densities J_p, J_n for U_a=-1 V

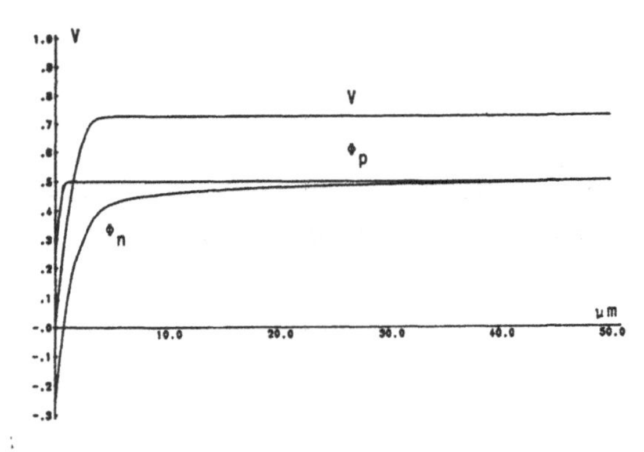

Fig.(4.4.10): Potentials V, Φ_p, Φ_n for U_a=-1 V

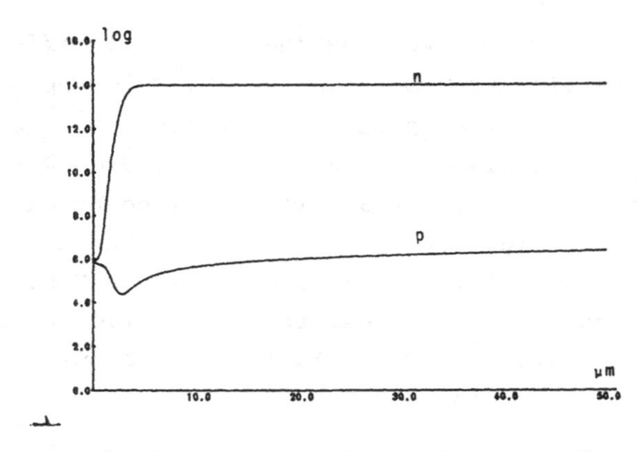

Fig.(4.4.11): Carrier densities p, n for U_a=-1 V

Summary

A collocation method using tension splines is given for the numerical solution of singularly perturbed boundary value problems. The strategy of choosing the tension parameter from the eigenvalues of the functional matrix of the differential equation system avoids oscillations in the numerical solution at the boundary layers. In all tested examples the algorithm performs successfully as regards both efficiency and accuracy.

Acknowledgement

I thank Professor Dr. R. Bulirsch for encouraging this work and Dr. P. Rentrop for helpful discussions.

References

1. Abrahamsson, L.: Numerical solution of a boundary value problem in semiconductor physics, Uppsala University Report No. 81 (1979).

2. Abrahamsson, L., Keller, H.B., Kreiss, H.O.: Difference approximations for singular perturbations of systems of ordinary differential equations, Numer. Math. 22, 367-391 (1974).

3. Ascher, U.: Solving boundary value problems with a spline-collocation code, J. Comput. Phys. 34, 401-413 (1980).

4. Ascher, U., Weiss, R.: Collocation for singular perturbation problems. First-order systems with constant coefficients, SIAM J. Num. 20 (3), 537-557 (1983).

5. de Boor, C.: Good approximation by splines with variable knots II. Lecture notes in Math., vol.363, Springer 1973

6. Broyden, C.G.: A class of methods for solving nonlinear simultaneous equations, Math. Comp. 19, 557-593 (1965).

7. Deuflhard, P.: A modified Newton method for the solution of ill-conditioned systems of nonlinear equations with application in multiple shooting, Numer. Math. 22, 289-315 (1974).

8. Deuflhard, P.: A relaxation strategy for the modified Newton method, Springer Lecture Notes, Optimization and Optimal Control (ed. Bulirsch, Oettli, Stoer) 477, 59-73 (1975).

9. Dickmanns, E.D., Well, K.H.: Approximate solution of optimal control problems using third order Hermite polynomial functions, Lecture Notes in Comp. Science 27, 158-166 (1975).

10. Flaherty, J.E., Mathon, W.: Collocation with polynomial and tension splines for singularly-perturbed boundary value problems, SIAM J. Sci. Stat. Comput. 1, 260-289 (1980).

11. Flaherty, J.E., O'Malley, R.E., Jr.: The numerical solution of boundary value problems for stiff differential equations, Math. Comp. 31, 66-93 (1977).

12. Howes, F.A.: Some old and new results on singularly perturbed boundary value problems, Singular Perturbations and Asymptotics (ed. Meyer, R.E., Parter, S.V.) Academic Press (1980).

13. Kreiss, H.O.: Difference methods for stiff ordinary differential equations, SIAM J. Numer. Anal. 15, 21-58 (1978).

14. Maier, M.R., Smith, D.R.: Numerical solution of a symmetric diode model, J. Comput. Phys. 42, 309-326 (1981).

15. Miranker, W.L.: Numerical methods of boundary layer type for stiff systems of differential equations, Computing 11, 221-234 (1973).

16. O'Malley, R.E., Jr.: Introduction to Singular Perturbations, Academic Press (1974).

17. Pearson, C.E.: On a differential equation of boundary layer type, J. Math. Phys. 47, 134-154 (1968).

18. Pearson, C.E.: On nonlinear differential equations of boundary layer type, J. Math. Phys. 47, 351-358 (1968).

19. Rentrop, P.: An algorithm for the computation of the exponential spline, Numer. Math. 35, 81-93 (1980).

20. Rieger, J.: Eindimensionale numerische Berechnung von physikalischen Grundgrößen einer pn-Diode mit beliebigem Dotierungsprofil. Technische Universität München, Institut für Technische Elektronik, Diplomarbeit 1981.

21. Russell, R.D.: Collocation for systems of boundary value problems, Numer. Math. 23, 119-133 (1974).

22. Russell, R.D., Christiansen, J.: Adaptive mesh selection strategies for solving boundary value problems, SIAM J. Numer. Anal. 15 (1), 59-80 (1978).

23. Sigfridsson, B., Lindström, J.L.: Electrical properties of electron-irradiated n-type silicon, J. Appl. Phys. 47, 4611-4620 (1976).

24. Smith, D.R.: On a singularly perturbed boundary value problem arising in the physical theory of semiconductors, Technische Universität München, Institut für Mathematik, TUM-M8021 (1980).

25. Stoer,J., Bulirsch, R.: Introduction to Numerical Analysis, Springer (1980).

26. Vasil'eva, A.B., Stel'makh, V.G.: Singularly disturbed systems of the theory of semiconductor devices, USSR Comp. Math. and Phys. 17, 48-58 (1977).

Progress in Scientific Computing, Vol. 5
Numerical Boundary Value ODEs

Solving Boundary Value Problems for Functional Differential Equations by Collation

G. Bader

1. Introduction

In many fields of application, such as Chemistry or Biology, processes appear which are more naturally modeled by functional differential equations (FDE's) than by ordinary differential equations (ODE's). Boundary value problems (BVP's) for FDE's appear frequently in the context of optimal control problems.

The numerical solution of problems involving FDE's has recently obtained much attention [5,7, 14, 15, 19, 20]. There are a number of different approaches, but it seems that the most popular one is the approximation through systems of ODE's, c.f. Banks/Kappel [7]. Later on in this section, the advantages and disadvantages of known approaches are briefly discussed.

This paper presents a new approach and a general purpose code for solving boundary value FDE's. A distinguishing feature is that possible nonsmooth behavior of the solution is taken into account. Our approach is based on a direct discretization of FDE's by collocation as outlined in Section 2. Results of a convergence analysis for nonlinear problems are given there as well. Section 3 gives a brief description of the code FDECOL, implemented on the basis of this approach. In the final section some numerical results are presented.

The type of problem treated is best described by an example (Thomas [20]). Consider the minimization of the functional

$$(1.1) \qquad I[u] = \int_1^2 u^2(t)dt$$

under the constraints

$$(1.2) \qquad \begin{aligned} y''(t) &= - y(t-1) + u(t), \quad t \in [0,2] \\ y(t) &= \phi(t) \qquad \text{for} \quad t \in [-1,0] \\ y'(0) &= \phi'(0) \\ y(2) &= y'(2) = 0 \end{aligned}$$

Application of a maximum principle for FDE's (see Jacobs/Kao [12], Banks [6]) yields a set of conditions for an optimal solution. For the given example we find

(1.3) $$y''(t) = -y(t-1) + u(t)$$
$$\lambda''(t) = \lambda(t+1), \quad \text{for} \quad t \in [0,2]$$

where $u(t) = -\lambda(t)/2$ and boundary conditions

(1.4) $$y(t) = \phi(t) \qquad \text{for} \quad t \in [-1, 0]$$
$$\lambda(t) = 0 \qquad \text{for} \quad t \in (2,3]$$
$$y'(0) = \phi'(0)$$
$$y(2) = y'(2) = 0.$$

Remark: Compared to optimal control problems for ODE's we have to treat three additional aspects:

i) the differential system contains both **retarded** and **advanced** arguments,

ii) parts of the boundary conditions are given in the form of initial - and end-functions,

iii) solutions are in general only piecewise smooth. Discontinuities may even occur at $t=a$ and at $t=b$.

Before we outline our approach, the advantages and disadvantages of known approaches are briefly discussed.

Transformation techniques: One of the earliest techniques treating difference - differential equations is the transformation into systems of ODE's. A description for retarded equations is found in Keller [13] and Ascher/Russell [4]. The generalization to problems with advanced arguments is straightforward. An advantage of this technique is that it allows the use of existing software for the solution of the transformed problems. On the other hand, the transformation exists only under rather special circumstances (c.f. [4]). Also, the number of ODE's is large for small delays and long integration intervals.

Approximation techniques: This approach uses systems of ODE's to approximate the solution of FDE's, see [7]. This approach is rather appealing because of its generality. From a more practical point of view, however, it has a number of inherent difficulties. It attempts to approximate only piecewise smooth solutions through globally smooth functions. An explicit treatment of nonsmooth solution behavior is not possible. As a consequence, the order of convergence is quite low. Finally the approximating system of ODE's is large and of singular perturbation type, hence a numerical solution is rather difficult.

This technique therefore seems not to be suited for practical purposes. See also section 4.

Direct solution through global methods: These methods are well known for the solution of BVP's in ODE's. For a collocation approach see deBoor/Swartz [9], Russell [17] and Ascher/Christiansen/Russell [2]. These methods solve nonlinear problems through linearization in function space. In solving FDE's this leads to well posed linear subproblems, i.e. linear BVP's, if the original BVP is well posed. This property is not shared by initial value methods, i.e. shooting, since the initial value problem for FDE's with advanced arguments is in general not well posed. As a consequence we treat only a collocation approach in the rest of this paper. If the nonsmooth solution behavior is explicitly treated then high orders of convergence can be retained, as we describe in the next section.

2. Collocation approach.

In order to keep the notation as transparent as possible, we restrict ourselves to first order systems with only one retarded and one advanced functional argument. A generalization to systems of higher order (see deBoor/Swartz [9]) as well as to a finite number of functional arguments is straightforward.

Consider a **nonlinear system** of FDE's

(2.1) $$y'(t) = F(t, y(t), y(\omega_1(t)), y(\omega_2(t)))$$

where $\omega_1(t) < t < \omega_2(t)$

on a finite interval $t \in [a,b]$ with **linear** (function) **boundary conditions**

(2.2) $$P_1 y(t) = \Phi(t) \quad \text{for} \quad t \in [\bar{a}, a)$$

$$P_2 y(t) = \phi(t) \quad \text{for} \quad t \in (b, \bar{b}]$$

and linear **side conditions**

(2.3) $$A_j y(\zeta_j) + B_j y(\omega_1(\zeta_j)) + C_j y(\omega_2(\zeta_j)) = \gamma_j$$
$$\text{for} \quad \zeta_j \in [a,b], \qquad 1 \leq j \leq n.$$

We assume that $y(t)$, $F(t,\cdot,\cdot,\cdot)$ and γ_j are vectors in $\mathbf{R}^n$, P_1, P_2 are projectors from $\mathbf{R}^n$ into $\mathbf{R}^n$ and $\omega_{1,2}(t)$ are continuous scalar functions on $[a,b]$ with $\min_{t \in [a,b]} \omega_1(t) = \bar{a}$, $\max_{t \in [a,b]} \omega_2(t) = \bar{b}$. A solution of the BVP (2.1) - (2.3) is in general only a piecewise smooth (vector-valued) function $y^*(t)$ for $t \in [\bar{a},\bar{b}]$. The solution may be discontinuous at $t=a$ or $t=b$ (see the above example).

Remark: For notational convenience we refer in the following to a solution y^* only as a function in $C[a,b]$.

As the BVP (2.1) - (2.3) only possesses piecewise smooth solution we formulate our approach in an appropriate space. Let

$$\pi: a = s_1 < s_2 < \ldots\ldots < s_{\hat{M}} < s_{\hat{M}+1} = b$$

be a partition of the interval $I = [a,b]$ and define

$$C_\pi[a,b] = C[s_1,s_2] \times C[s_2, s_3] \times \ldots\ldots \times C[s_{\hat{M}}, s_{\hat{M}+1}]$$

such that $f \in C_\pi[a,b]$ consists of $\hat{M}$ components $f_m \in C[s_m, s_{m+1}]$. If $f(t)$ is discontinuous for some $s_m \in \pi$, then let f be defined double valued by using its limiting values $f(s_m^-)$ and $f(_m^+)$. With the norm

$$\|f\| = \max_{1 \le m \le \hat{M}} \|f_m\|_\infty$$

$C_\pi[a,b]$ becomes a Banach space. Further, let $C_\pi^k[a,b]$ be the subspace of piecewise k-times continuously differentiable functions and $\mathcal{P}_{k,\pi}$ the subspace consisting componentwise of polynomials $p \in \mathcal{P}_k[s_m, s_{m+1}]$ for $1 \le m \le \hat{M}$.

Let the shift operators S_i: $C[a,b] \to C_\pi[a,b]$ be defined for $i = 1,2$ with $\pi \supset \pi_0 = \{\omega_1^{-1}(a), \omega_2^{-1}(b)\}$ as

$$S_i[u](t) := \begin{cases} \phi(\omega_i(t)) & \text{for } \omega_i(t) < a \\ u(\omega_i(t)) & \text{for } a \le \omega_i(t) \le b \\ \psi(\omega_i(t)) & \text{for } b < \omega_i(t) \end{cases}$$

(c.f. Reddien/Travis [16]). Using these operators the BVP (2.1) - (2.3) is more conveniently formulated for theoretical treatment as

$$(2.1') \qquad y'(t) = F(t, y(t), S_1[y](t), S_2[y](t))$$

and the side (boundary) conditions

$$(2.3') \qquad A_j y(\zeta_j) + B_j S_1[y](\zeta_j) + C_j S_2[y](\zeta_j) = \gamma_j$$
$$\zeta_j \in [a,b], \quad 1 \le j \le n \, .$$

After these preparations we state precisely the approach we use. Assume the right-hand side F of the FDE (2.1) is sufficiently smooth. Since the functional arguments $\omega_{1,2}(t)$ depend only on t we precompute the locations at which the solution and its derivatives up to order p has potential nonsmooth behavior. Including all these points in the partition π we have

$$y^* \in C_\pi^p[a,b] \cap C[a,b].$$

Remark: If we require locally more smoothness, i.e. increasing p, then the partition $\pi = \pi(p)$ becomes more dense in general. The appropriate choice of p for practical computation depends on the BVP and the required accuracy of the numerical solution.

To solve the BVP (2.1'), (2.3') numerically, a collocation method using continuous piecewise polynomials is applied. Thus, consider only partitions Δ

$$\Delta:\ a = t_1 < t_2 < \dots < t_M < t_{M+1} = b$$
$$h_i = t_{i+1} - t_i,\ h = \max_{1 \le i \le M} |h_i|$$

which include the partition $\pi (\pi \subset \Delta)$. On this partition (mesh) a k-stage collocation method is completely defined as a function of a fixed set of points

$$0 \le \rho_1 < \rho_2 < \dots < \rho_k \le 1$$

by requiring that the approximate solution y_Δ^* be componentwise in $C[a,b]$ and reduce to a polynomial of degree at most k on each subinterval $[t_i, t_{i+1}]$, that it satisfy the boundary conditions (2.3') and that it satisfy the differential equation (2.1') at the collocation points $t_{ij} = t_i + \rho_j h_i$, $1 \le i \le M$, $1 \le j \le k$ (c.f. deBoor/Swartz [9], Russell [17]).

Without giving details we summarize the basic convergence result from Bader [5] in the following theorem

Theorem 2.1:
Let $y^* \in C^{p+1}[a,b]$, $p \ge 1$, be a solution of the problem (2.1'), (2.3') and suppose

i) $F(t,y,S_1[y], S_2[y])$ is sufficiently smooth (e.g. $F \in C^2(N)$, N being a ε-neighborhood of the solution,

ii) the linearized problem associated with y^* is uniquely solvable and possesses a Green's function $H(t,s)$.

Then there exists ρ, ε so that

a) there is no other solution $\tilde{y}$ for $\|D(y^*-\tilde{y})\| < \varepsilon$

b) for $h \leq \rho$ there is a unique collocation solution $y^*_\Delta \in P_{k+1,\Delta} \cap C[a,b]$ in this neighborhood of y^*

c) Newton's method applied to the collocation equation converges quadratically in a neighborhood of y^*_Δ for $h \leq \rho$.

d) The following error estimates hold for $r = 0,1$

$$\|D^r(y^*-y^*_\Delta)\| \leq \text{const } h^{\min(p,k)}$$

e) Furthermore, for the collocation solution of the linearized problem z_Δ (c.f. [5], [9]) the estimates

$$\|D^r(y^*-z_\Delta)\| \leq \text{const } h^{\min(p,k)}$$

and

$$D^r(y^*-y^*_\Delta) = D^r(y^*-z_\Delta) + \theta(h^{2\min(p,k)})$$

hold.

Remark: Theorem 2.1 agrees formally with theorem 3.1 in [9] for the case of ODE's.

If we further assume sufficient smoothness of the Frechet derivative of the right hand side F of (2.1'), i.e. $DF(\cdot, y^*, S_1[y^*], S_2[y^*]) \in C^k_\pi[a,b]$, then convergence of higher derivatives of the collocation solution can be shown (c.f. [17]).

For ODE's, this is usually the departure point in deriving even higher order of convergence at mesh points, i.e. superconvergence for special sets of collocation points. A crucial basis for all these results is that the Green's matrix $H(t,s)$ has essentially the same smoothness as the solution itself when $t \neq s$. This property does not hold for FDE's, see [5] for a simple example. As a consequence superconvergence cannot be shown in general.

Remark: Under severe restriction of the class of problems considered and a special construction of the initial partition π superconvergence may be retained. Bellen [8] has analysed this for the special case of initial value problems.

But even if superconvergence is lost, the improvement of order of convergence over the approximation techniques is still substantial.

3. Aspects of Implementation

On the basis of the results of section 2, a code **FDECOL** **(Functional Differential Equations** treated by **COLlocation)** has been implemented. The implementation is based on the code COLSYS of Ascher/Christiansen/Russell [1,2,3,10,18]. The intention was to produce a general purpose code solving BVP's for FDE's. While the present code FDECOL already solves a large number of problems with success, further research is needed into error-estimation and - control to treat the "nonlocal behavior" of FDE's with more reliability.

To treat the class of problems considered in this paper, a large portion of the code COLSYS has been changed. In order to fit the scope of the present paper we restrict our description to the definition of the class of problems accepted by FDECOL and the solution of algebraic equations encountered for FDE's. A more complete description of the code is given in [5].

Problem definition:

Consider a **nonlinear** system of N mixed order FDE's

$$y_n^{(m_n)}(t) = F(t,z(t), Z(\omega(t)); par) \qquad (3.1)$$
$$a \le t \le b, \ 1 \le n \le N$$

where the sought solution $y(t) = (y_1(t),\ldots, y_N(t))$ is an isolated solution vector and

$$z(t) = z(y(t)) = (y_1,y_1,\ldots,y_1^{(m_1-1)}, y_2,\ldots,y_N^{(m_{N-1})})$$

is the vector of unknowns that would arise when converting (3.1) into a first order system. $\omega(t)$ is the vector of functional arguments $\omega(t) = (\omega_1,\ldots,\omega_L)$ with $\omega_\ell = \omega_\ell(t;par)$ and the parameter vector $par \in \mathbf{R}^P$. In the code we restrict $m_n \le 4$, $L \le 6$. The supervector $Z(\omega(t))$ of state variables is defined as $Z(\omega(t)) = (z(\omega_1(t)),\ldots,z(\omega_L(t)))$. The system (3.1) is subject to boundary conditions

$$P_1 z(t) = \phi(t) \qquad \text{for } t < a \qquad (3.2)$$
$$P_2 z(t) = \psi(t) \qquad \text{for } b < t$$

with appropriate projectors P_1, P_2 and **nonlinear** side conditions

$$g_n(\zeta_n;z(\zeta_n), Z(\omega(\zeta_n)); par) = 0 \qquad (3.3)$$
$$\zeta_j \in [a,b], \ 1 \le n \le \bar{m}$$

where $\bar{m} = m^* + p$, $m^* = \sum_{1}^{N} m_n$, $p = \dim(par)$.

For the numerical solution, let an initial partition (mesh) π be given. Assume the mesh π ensures that $y_n^* \in C_\pi^{k+m_n+r}[a,b]$ with $r > 0$ sufficiently large exists. We seek a numerical solution $v = (v_1,\ldots,v_N;\ par)$ where

$$(3.4) \qquad v_n \in P_{k+m_n,\Delta} \cap C^{(m_n-1)}[a,b],\ 1 \leq n \leq N$$

for meshes Δ with $\Delta \supset \pi$. As collocation points we use Gauss-Legendre points.

Solution of linear subproblems

Application of collocation to problem (3.1) - (3.3) leads, after linearization of the problem, to the solution of a sequence of linear problems. With $t_{i1},\ldots,t_{ik}$ the k Gaussian points in the ith subinterval $I_i = (t_i,\ t_{i+1})$, $1 \leq i \leq M$, write these problem as

$$(3.5) \qquad L_n[v]w(t_{ij}) = f(t_{ij})$$

$$1 \leq j \leq k,\ 1 \leq i \leq M,\ 1 \leq n \leq N$$

where L_n are the linearizations of the differential equations (3.1) and

$$(3.6) \qquad \beta_\ell[v]w = \gamma_\ell,\ 1 \leq \ell \leq m$$

the linearizations of the boundary conditions (3.3).
The number of equations in (3.5), (3.6) is $MNk + \bar{m}$, the dimension of the approximation space. Using B-splines as basis functions (see [3]) leads to a linear system of algebraic equations

$$(3.7) \qquad Ax = b$$

where the vectors of unknowns $x = (\alpha,\ par)$ contains the coefficients α with respect to the B-spline basis. The matrix A is determined through the linearized problem (3.5), (3.6) and the derivatives of the basis functions.

Consider next the structure of the matrix A. Fixing i and n, $1 < i < M$, $1 < n < N$, there are m_n nonzero B-splines on I_{i-1}, $k - m_n$ B-splines which vanish outside I_i and m_n which vanish outside $I_i \cup I_{i+1}$. Using an appropriate ordering of the coefficients $\alpha = (\alpha_{in})$ produces a block structured matrix A (see [1]).

Compared to the case of ordinary differential equations, additional off-diagonal blocks arise for FDE's. Their position and blocking depends dynamically on the mesh. The structure of the matrix

A is characterized follows:

Rows: With ℓ_i side conditions given at points ζ_j, $t_i \leq \zeta_j \leq t_{i+1}$, there are $kN + \ell_i$ corresponding rows. For each n, $1 \leq n \leq N$, k rows correspond to $t_{i1}, \ldots, t_{ik}$. The numbering of rows increases with the argument t.

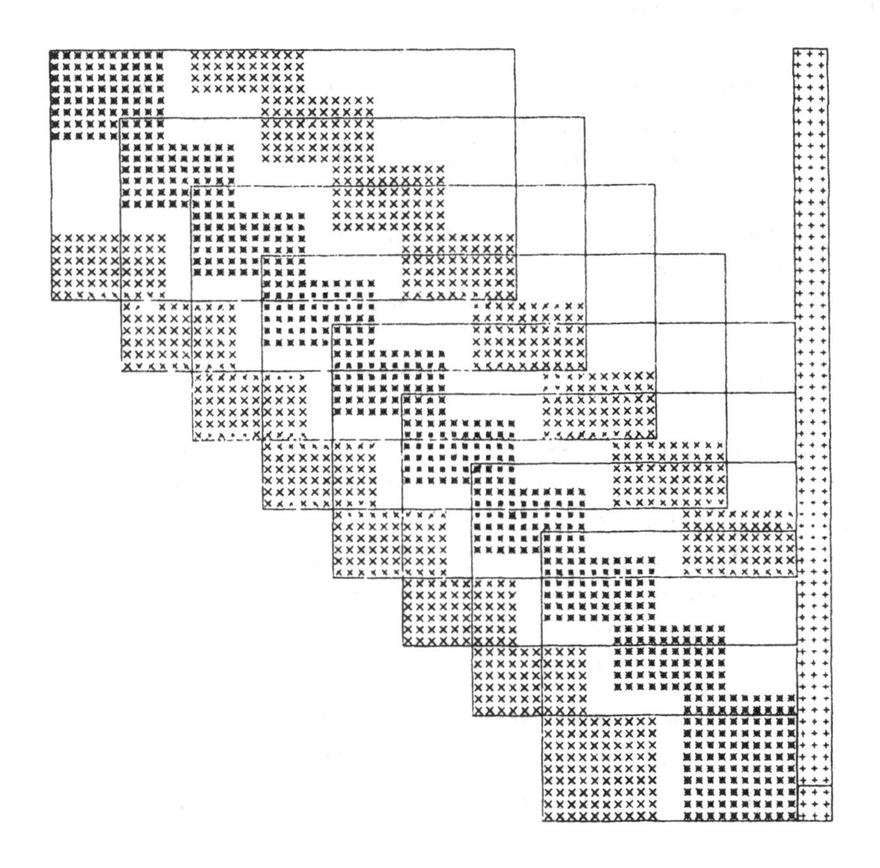

Fig. 3.1

Columns: i) central block: For each $v_n(\cdot)$, $1 < n < N$, there are m_n B-splines not vanishing on $I_i \cup I_{i-1}$. The corresponding columns, m_n for each n, will appear first in the order $m_1, m_2, \ldots, m_N$, totalling m^* columns. Then come $k-m_1$ $k-m_2, \ldots, k-m_N$ columns corresponding to $kN-m^*$ B-splines which vanish outside I_i. The m^* columns of those B splines which vanish outside $I_i \cup I_{i+1}$ appear last, ordered the same way as the first m^* columns. The total number of columns is therefore $kN+m^*$ plus the contribution from the parameters. ii) off diagonal blocks: For each functional argument ω_ℓ of (3.1) $Nk+m^*$ additional columns (in general off the diagonal) are produced. The ordering is as above. For $t = \omega_\ell$ (t_{ij}; par) $\in I_r$ the entries appear in the columns corresponding to $I_{r-1} \cup I_r \cup I_{r+1}$. For $I_r \neq \omega_\ell(I_i)$ a splitting of the off-diagonal blocks may arise.

Parameter block: Each parameter in the system (3.1) produces a corresponding column, placed in the right most columns of the matrix. An example with one retarded and one advanced argument containing 3 parameters is given in figure 3.1

The numerical solution of system (3.7) is done by means of so called frontal methods, see [5], also used in the finite element solution of partial differential equations.

4. Numerical results

The code FDECOL has been tested with a number of problems of different levels of complexity. A selection of examples may be found in [5]. For brevity, we present only three examples here, including a comparison with techniques from [7]. The computations were done in double precision on the IBM 370/768 at Heidelberg University.

In the examples below, the following notation is used:

$y_i(t)$ - ith component of the exact solution

tol$(y_i^{(j)})$ - absolute error tolerance for the component y_i (t)

est $E(y_i^{(j)})$ - estimated uniform error in $y_i^{(j)}(t)$

(FDECOL, like COLSYS, allows the user to specify different tolerances for different components.)

time - the actual solution time in seconds

$a \pm b$ - $a \cdot 10^{\pm b}$

k - number of collocation points per subinterval.
mesh sequence (iterations) - successive mesh sizes, i.e. numbers M of subintervals required, followed in parentheses by the numbers of Newton iterations performed on each mesh for nonlinear problems. Unless otherwise stated, the initial mesh for FDECOL is uniform.

Example 1: de Nevers/Schmitt [14].

We start with a standard test problem

$$(4.1) \qquad y''(t) = -\frac{1}{16}\sin y(t) - (t+1)\,y(t-1) + t$$

for $t \in [0,2]$ subject to boundary conditions

$$(4.2) \qquad y(t) = t - 1/2 \quad \text{for} \quad t \in [-1,0]$$

$$y(2) = -1/2.$$

Solutions to these examples are given in a number of papers using different types of approaches, see [4, 14, 20]. The solution y^* as a function in $C[0,1] \times C[1,2]$ is infinitely often differentiable - the only point to be fixed in partition is $t=1$. Results, starting with zero initial guess for the solution, are given in table 1.

Table 1

k	tol(y)	est E(y)	tol(y')	est E(y')	time	mesh sequence
3	1. -6	.5 -6	1. -5	.6 - 5	1.3	6(13), 12(1)

Remark: For the given mesh we observe superconvergence at the mesh - points. The reason is that the problem is from the restricted class mentioned at the end of section 2.

Example 2: Thomas [20]

This is an optimal control problem, having retarded arguments in the state and the control variable.

$$(4.3) \qquad y'(t) = -y(t) - y\left(t - \tfrac{1}{2}\right) - y(t-1) + u(t) + u\left(t - \tfrac{1}{2}\right)$$

$$y(t) \equiv 0 \quad \text{for} \quad t \in [-1,0].$$

A piecewise continuous control u is sought, which drives the state variable y under minimization of

(4.4) $$I[u] = \frac{1}{2} \int_0^{1.5} u^2(t)dt$$

into the final state $y(1.5) = 1$. For $t \in [-1/2, 0)$ $u_0(t) = -1$ is prescribed. Application of the maximum principle for FDE's yields the BVP

(4.5) $$\begin{aligned} y'(t) &= - y(t) - y(t - \tfrac{1}{2}) - y(t-1) + u(t) + u(t - \tfrac{1}{2}) \\ \lambda'(t) &= + \lambda(t) + \lambda(t + \tfrac{1}{2}) + \lambda(t+1) \end{aligned}$$

with the control $u(t) = -\left(\lambda(t) + \lambda(t + \frac{1}{2})\right)$ for $t \in [0,1.5]$ and the boundary conditions

(4.6) $$\begin{aligned} y(t) &= 0 && \text{for } t \in [-1,0] \\ u(t) &= -1 && \text{for } t \in [-\tfrac{1}{2}, 0) \\ \lambda(t) &= 0 && \text{for } t \in (1.5,2.5] \\ y(1.5) &= 1 . \end{aligned}$$

The solution $(y(t), \lambda(t))$ is infinitely often differentiable for $t \neq 0.5$ and $t \neq 1$. Results are tabulated in table 2.

Table 2

k	tol(y)	est E(y)	tol(λ)	est E(λ)	time	mesh sequence
4	1. -5	.3 -6	1. -5	.3 -6	0.85	3, 6

The functional value is $I[u] = 0.407409$ in agreement with Thomas [20]. A graph of the solution is given in figure 4.1 ($s(t) := u(t) + u(t - 1/2)$).

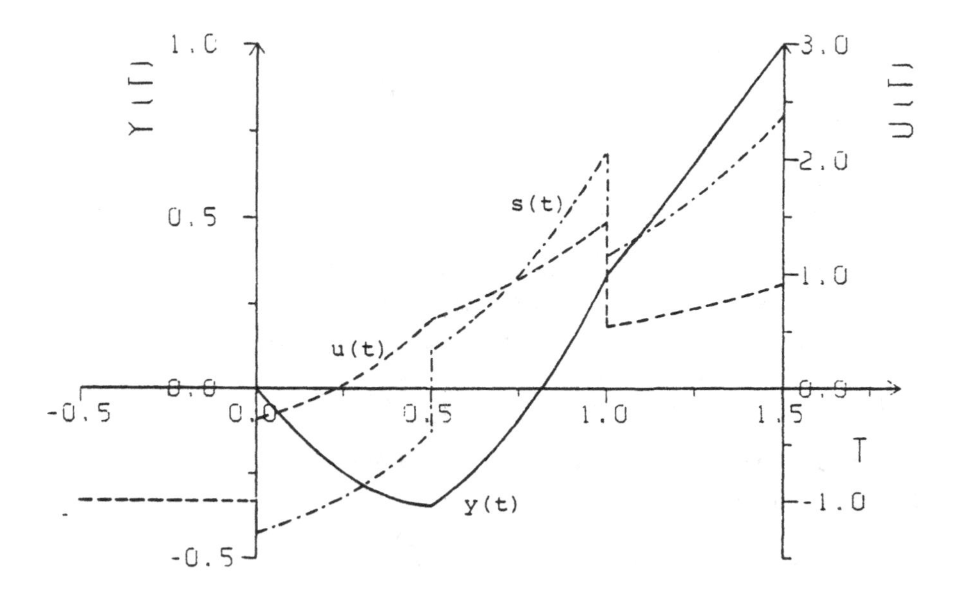

Fig 4.1

In [5], a comparison of the given results with approximations through systems of ODE's, discussed in section 1, was carried out. For moderate systems of approximating ODE's (n=13 equations) the obtained accuracy is rather poor. In the above example est E(y) = 0.12. This result might be expected from figure 4.1. For increasing size of the ODE system (n=25) results are only slightly more accurate at a drastically increased computing time for the numerical solution. We conclude that approximation techniques are not competative for the class of problems treated herein.

Example 3: Soliman/Ray [15, 19]
This is a much more challenging example modelling a continuously stirred tank reactor in which an exothermic, catalytic irreversible reaction is carred out. See Soliman/Ray [15, 19] for a description of the model and an approximate solution. The dynamical behavior of this reactor can be described by the following equations:

$$(4.7)\qquad \begin{aligned} y_1' &= -y_1 - BR \\ y_2' &= -y_2 + 0.9u_2^- + 0.1u_2 \\ y_3' &= -2y_3 + 0.25BR - u_1 y_3^-(y_3 + 0.125) \end{aligned}$$

where $y_i = y_i(t)$, $i=1,2,3$; $u_i = u_i(t)$, $i=1,2$

$$u_2^- = u_2(t-\tau_2),\ y_3^- = y_3(t-\tau_1)$$

$$BR = \left[(1+y_1)(1+y_2)\exp\left(\frac{25y_3}{1+y_3}\right) - 1\right].$$

The linear control u_1 is piecewise continuous with $|u_1| < 500$, $\tau_1 = \tau_2 = 0.015$. We attempt to drive the system (4.7) from some initial value into steady state by minimizing the functional

$$(4.8)\qquad I[u] = \int_0^{0.2} [y_1^2 + y_2^2 + y_3^2 + 0.01\, u_2^2](t)\, dt\,.$$

Applying a maximum principle [6,12] yields a set of 3 equations for the adjoint variables

$$(4.9)\qquad \begin{aligned} \lambda_1' &= -2y_1 + \lambda_1 + (\lambda_1 - 0.25\lambda_3)\frac{\partial BR}{\partial y_1} \\ \lambda_2' &= -2y_2 + \lambda_2 + (\lambda_1 - 0.25\lambda_3)\frac{\partial BR}{\partial y_2} \\ \lambda_3' &= -2y_3 + 2\lambda_3 + (\lambda_1 - 0.25\lambda_3)\frac{\partial BR}{\partial y_3} \\ &\quad + \lambda_3 u_1 y_3^- + \lambda_3^+ u_1^+ (y_3^+ + 0.125) \end{aligned}$$

where $\lambda_i = \lambda_i(t)$, $i = 1,2,3$, $\lambda_3^+ = \lambda_3(t+\tau_1)$, $y_3^+ = y_3(t+\tau_1)$, and the control variables

$$(4.10)\qquad \begin{aligned} u_1(t) &= 500\ \mathrm{sign}(S(t)) \\ S(t) &= \lambda_3 y_3^-(y_3 + 0.125)(t) \\ u_2(t) &= -(5\lambda_2 + 45\lambda_2^+)(t). \end{aligned}$$

The control $u_1(t)$ is a step function (**bang-bang** control) with jumps determined through the zeros of $S(t)$. As a consequence the jumps of $u_1(t)$ from its upper to its lower bound and vice versa depend on the solution. This and the strong nonlinearity of the equations (4.7) - (4.10) make the problem challenging.

Taking into account the fact that neither an initial guess for the adjoint variables (4.9) nor the switching structure for u_1 is known, a direct solution of the problem cannot be found. Instead, in a first step, an approximating problem is solved. For that purpose we fix the bang-bang control u_1 and relax the final conditions in a least square sense.

Since no better initial guess for the solution is available $(y_0(t), \lambda_0(t)) \equiv 0$ is used in the computations. To avoid convergence difficulties for Newton's method a continuation method, see Deuflhard/Pesch/Rentrop [11], is applied. Instead of giving all details, the results of our computations are presented in table 3 and figure 4.2.

Table 3

k	$to\ell(y_i)$	est $E(y_i)$	$to\ell(\lambda_i)$	est $E(\lambda_i)$	time	mesh sequence
2	1. -2	< 1. -3	1. -2	< 1. -2	29.5	20(15) 40(3) not aquidistant

The solution of the simplified problem provides now a good initial guess for the treatment of the complete optimal control problem (4.7), (4.8). This problem is still under investigation. Results will be published elsewhere.

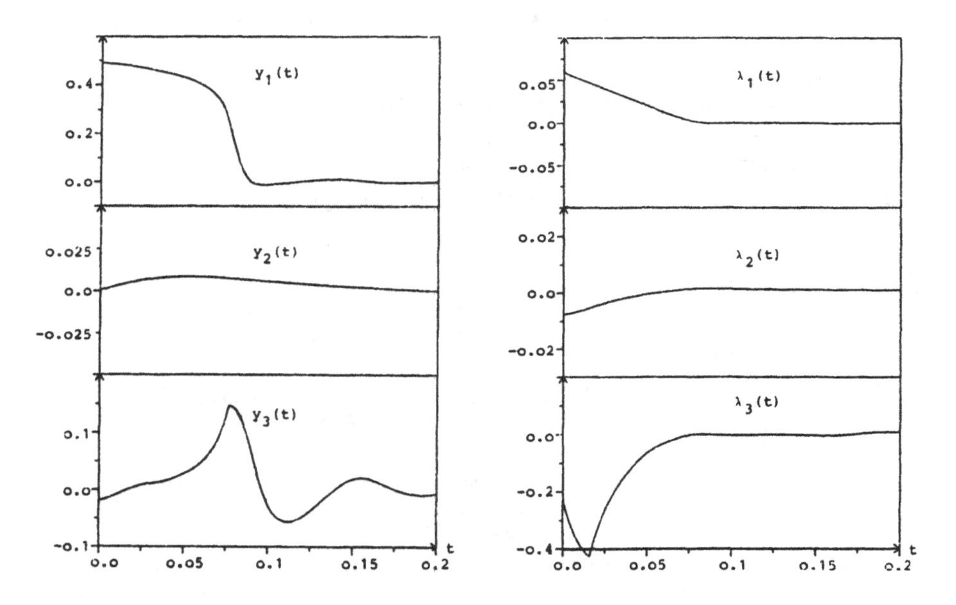

Fig. 4.2

Acknowledgements: The author would like to thank P. Deuflhard who encouraged this work which is part of the author's dissertation. This research was supported in part by the Deutsche Forschungsgeimeinschaft and by the National Science and Engineering Research Council of Canada.

References

[1] U. Ascher, J. Christiansen, R.D. Russell, "A Collocation Solver for Mixed Order Systems of Boundary Value Problems", Math. of Comp. vol. 33, 659-679, (1979).

[2] U. Ascher, J. Christiansen, R.D. Russell, "COLSYS A Collocation Code for Boundary Value Problems", Proc. Conf. for Codes for BVP's Houston, Texas, 1978.

[3] U. Ascher, R.D. Russell, "Evaluation of B-Splines for Solving Systems of Boundary Value Problems", Tech. Rep. 77-14, Dept. Computer Sc., University of B.C. Vancouver, Canada.

[4] U. Ascher, R.D. Russell, "Reformulation of Boundary Value Problems into Standard Form". SIAM Review vol. 23, 238-254 (1981).

[5] G. Bader,"Numerische Behandlung von Randwertproblemen für Funktional-Differentialgleichungen", University of Heidelberg, Ph.D. Thesis (1984).

[6] H.T. Banks, "Necessary Conditions for Control Problems with variable time Lags", SIAM J. Control, vol. 1, 9-47, (1968).

[7] H.T. Banks, F. Kappel, "Spline Approximations for Functional Differential Equations", J. of Diff. Equations, vol. 34, 496-522 (1979).

[8] A. Bellen, "One step Collocation for Delay Differential Equations", J. of Comp. Appl. Math., vol. 10, (1984).

[9] C. de Boor, B. Swartz, "Collocation at Gaussian Points", SIAM J. Numer. Anal., vol. 10, 582-606, (1973).

[10] J. Christiansen, R.D. Russell, "Error Analysis for Spline Collocation Methods with Application to Knot Selection" Math. of Comp., vol. 32, 415-419, (1978).

[11] P. Deuflhard, H.-J. Pesch, P. Rentrop, "A Modified Continuation Method for the Numerical Solution of Two-Point Boundary Value problems by Shooting Techniques", Numer. Math., vol. 26, 327-343, (1976).

[12] M.Q. Jacobs, Ti-Jeun Kao, "An Optimum Settling Problem for Time Lag Systems", J. of Math. Anal. and Appl., vol. 40, 687-707, (1972).

[13] H.B. Keller, "Accurate Difference Methods for linear ordinary Differential Systems subject to linear Constraints". SIAM J. Numer. Anal. vol. 6, 8-30, (1969).

[14] K. de Nevers, K. Schmitt, "An Application of the Shooting Method to Boundary Value Problems for Second Order Delay Equations". J. of Math. Anal. and Appl., vol. 36, 588-597 (1971).

[15] W.H. Ray, M.A. Soliman, "The Optimal Control of Processes containing pure Time Delays-I, Necessary Conditions for an Optimum." Chem. Engng. Sci., vol. 25, 1911-1925, (1970).

[16] G.W. Reddien, C.C. Travis, "Approximation Methods for Boundary Value Problems of Differential Equations with Functional Arguments", J. of Math. Analysis and Appl. vol. 46, 62-74, (1974).

[17] R.D. Russell, "Collocation for Systems of Boundary Value Problems", Num. Math. vol. 23, 119-133, 1974.

[18] R.D. Russell, J. Christiansen, "Adaptive Mesh Selection Strategies for Solving Boundary Value Problems", SIAM J. Numer. Anal. vol. 15, 59-80, (1978).

[19] M.A. Soliman, W.H. Ray, "Optimal Control of Multivariable Systems with Pure Time Delays", Automatica vol. 7, 681-689, (1971).

[20] B. Thomas, "Numerische Behandlung von retardierten Differentialgleichungen mit Hilfe der Extrapolationsmethode und Anwendungen auf retardierte Randwertprobleme." Universität Köln, Diplomarbeit 1973/1974.

Progress in Scientific Computing, Vol. 5
Numerical Boundary Value ODEs

THE APPROXIMATION OF SIMPLE SINGULARITIES

A. Griewank
G. W. Reddien

1. Introduction

Many physical problems can be formulated as a parameter dependent nonlinear operator equation

$$f(z,\lambda) = 0, \quad f : D \subset Z \times R \to Y \tag{1.1}$$

where Z and Y are Banach spaces and D is an open set in $Z \times R$. Even if the original problem does not contain any explicit "control" parameters such as λ in (1.1), it may be advantageous to relax some physical constant, e.g., viscosity, or introduce an artificial parameter. By then analyzing (1.1) for certain critical or special values of λ, one may be able to characterize the full solution set

$$M_\lambda \equiv \{z \in Z : f(z,\lambda) = 0\}$$

for any given value of λ. Whenever the Frechet derivative f_z has range all of Y, the slices M_λ form locally smooth manifolds of dimension

$$i \equiv \text{null}(f_z) - \text{def}(f_z)$$

and vary smoothly in λ. Here i denotes the general Fredholm index of f_z, which we assume to be constant and nonnegative in the domain D. Of particular interest is the square case $i = 0$, in which the solutions $z = z(\lambda) \in M_\lambda$ are locally unique and differentiable in λ as long as $f_z(z(\lambda),\lambda)$ possesses a bounded inverse. This observation is the basis of continuation or homotopy methods [14].

The geometrical structure of M_λ may undergo drastic changes at points $(z_0,\lambda_0) \in f^{-1}(0)$ where the range of $f_z^0 \equiv f_z(z_0,\lambda_0)$ is a proper subspace of Y. There are many different types of such singular solutions (z_0,λ_0), and the resulting shape of the critical slice M_{λ_0} may be very complicated. It is reasonable to ask whether or not it is useful to compute such complicated singularities. Usually a dynamical system which has (1.1) as its steady state equation exhibits a jump to a different "branch" of solutions only near simple turning points. This would be the square case $i = 0$ with

$$\operatorname{def}(f_z^0) = 1, \operatorname{null}(f_z^0) = 1 \text{ and } \operatorname{def}(f_z^0, f_\lambda^0) = 0 . \tag{1.2}$$

However, it has been profitable to view "higher" singularities as "organization centers" for solution diagrams of steady states. See, for example, Golubitsky and Keyfitz [11] and also Beyn and Bohl [3].

In this paper we will restrict our attention to singularities of corank one as in (1.2) and will allow $\operatorname{null}(f_z^0) = i+1$. The condition $\operatorname{def}(f_z^0, f_\lambda^0) = 0$ means then that the set $f^{-1}(0) \subset D$ forms a smooth manifold of dimension $i+1$ near (z_0, λ_0). This condition is the only requirement on the way in which the parameter λ enters into the operator $f(z,\lambda)$. In the finite dimensional case $n = \dim(Z) < \infty$, the conditions (1.2) require that the $(n-i) \times n$ matrix f_z^0 drops in rank by one and that the n-vector f_λ^0 is not in range (f_z^0).

The singularities may be characterized theoretically by first reducing the number of variables. This reduction can be accomplished by the implicit function theorem or the Liapunov-Schmidt procedure. In the finite dimensional case, for example, we eliminate $n-i-1$ equations and variables and express λ as a differentiable function $\phi(\bar{z})$ of the remaining variables $\bar{z} \in R^{i+1}$. A similar reduction to a function ϕ on R^{i+1} occurs in the infinite dimensional case. By construction, the slices M_λ are topologically equivalent to the level sets $\phi^{-1}(\lambda)$ near the stationary point $\bar{z}_0$ where $\nabla\phi(\bar{z}_0) = 0$. If the Hessian $H_0 \equiv \nabla^2\phi(\bar{z}_0)$ is nonsingular, the geometrical structure of $\phi^{-1}(\lambda)$ is determined by its Morse index [10]. The most important cases - index zero or one - will be discussed briefly below.

If ϕ is degenerate in that $\det(H_0) = 0$, one has to take into account the cubic terms in the Taylor expansion of ϕ at $\bar{z}_0$ in order to develop the appropriate algebraic bifurcation equations. These equations are much harder to analyze. Such higher order singularities are likely to disappear if the original operator $f(z,\lambda)$ is discretized or otherwise perturbed. To make them generic, one has to introduce at least one additional parameter. The minimal number of parameters required to ensure the existence of a certain type of singularity for all sufficiently small smooth perturbations is called its codimension [10]. In this paper we consider only singularities of codimension 1, i.e., we assume throughout that $\det(H_0) \neq 0$.

In the simple (or quadratic) turning point case (1.2), the "matrix" H_0 reduces to a single scalar which can be expressed as

$$H_0 = u_0^* f_{zz}^0 v_0 v_0$$

for suitable vectors $v_0 \in Z$ and $u_0^* \in Z^*$ which span the nullspaces of f_z^0 and $(f_z^0)^*$, respectively. The leading terms in the reduced problem $\lambda - \phi(\bar{z}) = 0$ take the simple form

$$\lambda - \lambda_0 = \frac{1}{2}(\bar{z} - \bar{z}_0)^T H_0 (\bar{z} - \bar{z}_0). \tag{1.3}$$

Thus for λ near λ_0, the level sets $\phi^{-1}(\lambda)$ and the slices M_λ are either empty or consist of two nearby points if $(\lambda - \lambda_0)H_0$ is nonzero. Therefore, any attempt to solve $f(z,\lambda)$ for fixed λ near λ_0 must either fail or lead to serious numerical difficulties due to the singularity or near singularity of f_z. However, since $f^{-1}(0)$ forms locally a smooth curve, it can be traced in terms of a parameter such as pseudo-arclength [14].

In some applications, the exact computation of λ_0, which could represent a critical load or temperature, is of practical importance. To this end the underdetermined system $f(z,\lambda) = 0$ can be appended by an equation $g(z,\lambda) = 0$ that enforces the singularity of f_z. For $i = 0$, the nonsingular roots of the defining system

$$f(z,\lambda) = 0, \quad g(z,\lambda) = 0 \tag{1.4}$$

will be exactly the quadratic turning points of f. The main difficulty in solving (1.4) by Newton's method is that the gradient of g, however g is defined, depends on second derivatives of f which may be costly to evaluate.

If $i = 1$, the symmetric 2×2 matrix H_0 can be expressed as

$$H_0 \equiv \begin{pmatrix} u_0^* f_{zz}^0 v_0 v_0 & u_0^* f_{zz}^0 v_0 w_0 \\ u_0^* f_{zz}^0 v_0 w_0 & u_0^* f_{zz}^0 w_0 w_0 \end{pmatrix} \tag{1.5}$$

where $f_{zz}^0 \equiv f_{zz}(z_0,\lambda_0)$ and the vectors $u_0^* \in Z^*$ and $v_0, w_0 \in Z$ span the nullspaces of $(f_z^0)^*$ and f_z^0, respectively. If $\det(H_0) > 0$, we may see from (1.3) that both $\phi^{-1}(\lambda_0)$ and M_{λ_0} consist of a single isola formation point. For $\lambda \approx \lambda_0$, the sets $\phi^{-1}(\lambda)$ and M_λ are either empty or form a closed loop if $(\lambda - \lambda_0)H_0$ is positive definite or negative definite, respectively. If the Crandall-Rabinowitz condition $\det(H_0) < 0$ [6] is satisfied, simple bifurcation occurs--i.e., M_{λ_0}, like $\phi^{-1}(0)$, consists of two smooth curves through z_0.

Complications arise if in an infinite dimensional setting f is replaced by a discretization f^n. In the turning point case, f^n will also have turning points (z_n,λ_n), which converge to (z_0,λ_0) as $n \to \infty$. Often superconvergence occurs for $\lambda_n - \lambda_0$ [5]. In the bifurcation case, the discrete solution sets $\{f^n(z,\lambda_0) = 0\}$, like the slices M_λ for $\lambda \neq \lambda_0$, consist of two hyperbolic curves that come close, but do not intersect. The maximal distance between nearest points in $\{f^n(z,\lambda_0) = 0\}$ and M_{λ_0} declines only like the square root of the unavoidable discretization errors [5]. One might prefer to use a defining system of the form (1.4) with $g(z,\lambda)$ now a two-component function. (An explicit procedure to find g will be given in Section 2.) Having computed a root (z_n,λ_n) of the corresponding discretized equation, one may then determine the nature of (z_0,λ_0) by checking the sign of $\det(H_0^n)$, where H_0^n is defined by (1.5) with f replaced by f^n. If $\det(H_0^n) < 0$, the slice $\{z : f^n(z,\lambda_n) = 0\}$ consists locally of two intersecting branches, $z^n(\tau)$ and $y^n(\tau)$. Moreover, it can be seen that for suitable parametrizations, the distances $\|z^n(\tau) - z(\tau)\|$ and $\|y^n(\tau) - y(\tau)\|$ are of the same order as the discretization error. Thus the direct computation of approximations (z_n,λ_n) to (z_0,λ_0) yields an accurate representation of isola formation or simple bifurcation phenomena.

The main concerns of this paper are the definition and relation between the functions $g(z,\lambda)$ and their discretizations $g^n(z,\lambda)$ as well as the resulting errors between (z_n,λ_n) and (z_0,λ_0). The analysis is developed for an even larger class of points with rank drop one, which allows us, for example, to exploit the special structure of symmetry breaking bifurcations [22].

The second section of this paper contains a description of our characterization of singular points. The characterization will be without initial reference to $f(z,\lambda)$. Rather we will characterize the set of singular points for a general parameter dependent family, $B(x)$, of linear maps. This family does not necessarily arise directly as the derivative of f. As just described, an elimination of variables or reduction via symmetry for f may take place prior to the characterization of singularities.

In section 3 we give our approximation results, and in section 4 we give applications in the setting of two-point boundary value problems. Other characterizations in addition to those already mentioned can be found in [12], [16], [17], [18] and [20]. The characterization given in [12] in a finite dimensional setting has several advantages. It provides a single mathematical framework for the computation of singular points defined by the conditions (1.2), it requires the minimal number of equations in (1.4), and it includes explicit formulas for the needed derivatives in order to solve (1.4) by Newton's method. This procedure was extended to an infinite dimensional setting with a few approximation results [13]. The results presented here generalize and improve those of [13]. The setting here is more general, and sharper error estimates for the case of discretization by projection methods in the context of two-point boundary value problems are derived.

2. Characterization of Simple Singularities

Let X, Z_1 and Z_2 be Banach spaces and let D be an open set in X. Let $B(x)$ be a completely continuous linear operator from Z_1 to Z_2 and continuously differentiable for x in D. Let $\operatorname{ind}(B(x)) = i \geq 0$ on D, and assume also that $\operatorname{def}(B(x)) \leq 1$. In the finite dimensional case, $B(x)$ would be an $m \times n$ matrix with $i = n - m$. As applied to (1.1), $B(x)$ would be F_z with $x = (z,\lambda)$.

The singular set S will be defined by

$$S = \{x \in D : \operatorname{def}(B(x)) = 1\} ,$$

and we next want to characterize S for computational purposes. Choose $r(x)$ in Z_2 and $T(x)$ in $[Z_1, R^{i+1}]$ to be completely continuous and continuously differentiable in x such that for a given x_0 in S, $r_0 \equiv r(x_0)$ is not in range $(B(x_0))$, and $T(x_0)w = 0$, $B(x_0)w = 0$ imply $w = 0$. As an example, let the matrix operator B have the form

$$\begin{bmatrix} B_1 & B_2 \\ b_1^T & b_2^T \end{bmatrix}$$

with $b_1^T(x_0) = 0$, $b_2^T(x_0) = 0$ and $\det(B_1(x_0)) \neq 0$. The vector b_2 is in R^{i+1}. Then one could choose $r = [0,1]^T$ and $T = [0, I_{i+1}]$ in order to satisfy these conditions. In any case it follows from our conditions on r and T that

<u>Lemma 2.1</u>. The linear operator $A(x) \equiv \begin{bmatrix} B(x) & r(x) \\ T(x) & 0 \end{bmatrix} : Z_1 \times R \to Z_2 \times R^{i+1}$

has a bounded inverse for x in some neighborhood of x_0. Moreover there exist smooth functions $V(x) \in Z_1^{i+1}$, $g(x) \in R^{i+1}$ and $u^*(x)$ in Z_2^* satisfying

$$\begin{aligned} BV &= rg^T \qquad & u^*B &= g^TT \\ TV &= I_{i+1} \qquad & u^*r &= 1 \end{aligned} \tag{2.1}$$

It follows from Lemma 2.1 that x is in S, the singular set, if and only if $g(x) = 0$. Associated null vectors are given by the columns of V. It also follows that

$$g(x)^T = u^*(x)B(x)V(x). \tag{2.2}$$

The formulas (2.1) and (2.2) also lead to the result

$$g'(x)^T = u^*(B'V - r'g^T) - g^TT'V \tag{2.3}$$

where the prime denotes differentiation with respect to x. If x is in S or if r and T are constants, then the formula simplifies to

$$g'(x)^T = u^*(x)B'(x)V(x). \tag{2.4}$$

Each component g_j of g satisfies the relationship

$$g_j'(x) = u^*(x)B'(x)v_j(x) . \tag{2.5}$$

If $B(x) = f_z(z_0,x)$ for some additional function f and fixed z_0, then (2.5) leads to the approximation formula

$$\begin{aligned} g_j'(x) &= u^*(x)f_z'(z_0,x)v_j(x) \\ &\approx u^*(x)[f'(z_0 + \varepsilon v_j(x),x) - f'(z_0,x)]/\varepsilon . \end{aligned} \tag{2.6}$$

The formula in (2.6) allows the approximation of $g'(x)$ with $i+1$ additional evaluations of f'. Direct approximation of f_z' by differencing in every component of x could be expensive. We also make the nondegeneracy assumption $\operatorname{def}(g'(x)) = 0$. This condition implies that S will be a smooth manifold of codimension $i+1$.

Now suppose we have a second nonlinear Fredholm operator $f : X \to Y$, where Y is a Banach space such that on D, $\operatorname{ind}(f') = \operatorname{null}(f') = i+1$. Then the solution of the appended system

$$\begin{aligned} f(x) &= 0 \\ g(x) &= 0 \end{aligned} \tag{2.7}$$

produces intersections of the manifolds S and $f^{-1}(0)$. We assume $f(x_0) = 0$ for some x_0 in S. The system (2.7) will be regular at such an intersection point if these two manifolds are transversal--i.e., they have no common tangent. But this will be the case if and only if the $(i+1) \times (i+1)$ matrix $H \equiv g'(x_0)W$ is nonsingular where $\operatorname{kern}(f'(x_0)) = \operatorname{span}[w_1, w_2, \ldots, w_{i+1}]$. Using (2.4), the matrix H can be written as

$$H = (u_0^* B_0' w_r w_s)_{\substack{r=1,\ldots,i+1 \\ s=1,\ldots,i+1}} \tag{2.8}$$

This result should be compared to (1.5).

Let $X = Z \times R$ so that our problem may be written in the form $f(z,\lambda)$ as in (1.1). Let $Z_1 = Z$ and $Z_2 = Y$ where $f : X \to Y$ and define $B(x) = f_z$. Next assume that at x_0 in S, conditions (1.2) hold. The symmetric matrix H in (2.8) becomes

$$H = (u_0^* f_{zz}^0 v_r v_s)_{\substack{r=1,\ldots,i+1 \\ s=1,\ldots,i+1}}$$

where $f_z^0 V = 0$ and $TV = I_{i+1}$. Since f_λ is not in the range of B_0, then the solution manifold, $f^{-1}(0)$, must be "perpendicular" to the λ-axis, i.e., the tangent space to $f^{-1}(0)$ at x_0 is spanned by V. Thus such points are called generalized turning points for f [12]. As mentioned in the introduction, if $i = 0$ and $H \neq 0$, then x_0 is a quadratic turning point for f. In the case $i = 1$, $\det(H) < 0$ implies simple bifurcation occurs at x_0 [6]. If $\det(H) > 0$, then x_0 is an isola formation point [7].

We next consider symmetry breaking bifurcation. This case will use the generality of our characterization of singularities. Let $F(w,\lambda)$ map $W \times R$ into W where W is a Banach space. We assume the common symmetry condition $F(Qw,\lambda) = QF(w,\lambda)$ where $Q \neq I$ is a bounded linear map satisfying $Q^2 = I$. The Banach space W may be decomposed into its symmetric and antisymmetric elements-- $W = W_s \oplus W_a$ where $W_s = \frac{1}{2}(I - Q)W$ and $W_a = \frac{1}{2}(I + Q)W$. It follows directly that F maps $W \times R$ into W_s, and for $\hat{w}$ in W_s, $F_w(\hat{w},\lambda)$ maps W_s into W_s and W_a into W_a. Then for symmetry breaking bifurcation, a solution branch for $F = 0$ lies in $W_s \times R$ and at a bifurcation point x_0, the tangent to the bifurcating branch lies in W_a. To put this problem in our general framework, we let $X = W_s \times R$, $Z_1 = W_a$, $f(x) = F|_X$ and $B(x) = F'|_{W_a}$. Then the usual symmetry breaking bifurcation conditions [22] imply $\text{ind}(f(x)) = 1$ and $\text{ind}(B(x)) = i = 0$. The iterative computation of $x_0 \in W_s \times R$ by solving (2.7) with Newton's method requires one evaluation of the full Jacobian $F'(x)$ and one of its antisymmetric part $B(x)$ per step.

3. Discretization by Projection Methods

In this section, we consider the approximation of the system (2.7) through the use of projection methods [19]. Other authors have considered the approximation of solution manifolds via general discretizations, notably Fink and Rheinboldt [9] and Brezzi, Rappaz and Raviart [5]. Our interest here is not in the approximation of the solution manifold, but rather in the direct approximation of the singular point. Our results are more general than [5] in that, for example, generalized turning points are allowed. We also extend our analysis to derive superconvergence results in the case of collocation at Gauss points when applied to two-point boundary value problems. Similar results are obtained for the Galerkin method and turning points in [5].

We assume that $B(x) = P + R(x)$ where P is a linear projection mapping Z into PZ with nullity i. Here $Z = Z_1$ and $PZ = Z_2$. We assume $R(x)$ is a completely continuous mapping from Z into PZ and continuously differentiable in x. Let Z_n be a finite dimensional subspace of Z and let $\{\lambda_j\}_{j=1}^n$ be continuous linear functionals in $(PZ)^*$. Let $(I - P)Z \subset Z_n$ and let $\dim(Z_n) = n + i$. Then a well-defined projection, P_n, mapping Z into Z_n is given by $P_n z = z_n$ if and only if (i) $z_n \in Z_n$, (ii) $\lambda_j(Pz) = \lambda_j(Pz_n)$, $j = 1,\ldots,n$, and (iii) $(I - P)z = (I - P)z_n$, provided that $\lambda_j(Pz_n) = 0$, $j = 1,\ldots,n$, implies $Pz_n = 0$. One should think of Z as the cartesian product of a Banach space with an i-dimensional space of control parameters. The parameters are left alone by the projection. We assume $\lim_{n\to\infty} P_n z = z$. Define $B_n(x) = P + P_n R(x)$ and

$$A_n(x) = \begin{bmatrix} B_n(x) , & P_n r(x) \\ T(x) , & 0 \end{bmatrix} .$$

Then the following lemma is straightforward.

Lemma 3.1. There exists a neighborhood Ω of x_0 so that $A_n(x) \to A(x)$ uniformly on Ω as $n \to \infty$.

As an immediate consequence of this lemma, it follows that A_n^{-1} exists for n sufficiently large and also that $A_n^{-1} \to A^{-1}$ uniformly on Ω as $n \to \infty$. We next approximate the system (2.1) by solving

$$\begin{aligned} B_n V_n &= (P_n r) g^T & \qquad u_n^* B_n &= g_n^T T \\ TV_n &= I_{i+1} & \qquad u_n^*(P_n r) &= 1 \end{aligned} \tag{3.1}$$

where V_n is in Z_n^{i+1} and g_n is in R^{i+1}. The two sets of equations in (3.1) may be written as

$$A_n \begin{pmatrix} V_n \\ -g_n^T \end{pmatrix} = \begin{pmatrix} 0 \\ I_{i+1} \end{pmatrix}, \quad A_n^* \begin{pmatrix} u_n \\ -g_n \end{pmatrix} = \begin{pmatrix} 0 \\ 1 \end{pmatrix}.$$

The functional u_n^* does usually not belong to $P_n^*(PZ)^*$. However, since $PV_n = P_n PV_n$ we have

$$g_n^T = u_n^*(P + P_n R(x))V_n = (P_n^* u_n^*)(P + P_n R(x))V_n,$$

and correspondingly

$$(g_n')^T = (P_n^* u_n^*)[R'(x)V_n - r' g_n^T] - g_n^T T' V_n.$$

Consequently, only the projection $P_n^* u_n^*$ is needed to actually evaluate g_n and g_n'. To obtain $P_n^* u_n^*$ together with V_n and g_n we solve the systems

$$\tilde{A}_n \begin{pmatrix} V_n \\ -g_n^T \end{pmatrix} = \begin{pmatrix} 0 \\ I_{i+1} \end{pmatrix}, \quad \tilde{A}_n^* \begin{pmatrix} P_n^* u_n^* \\ -g_n \end{pmatrix} = \begin{pmatrix} 0 \\ 1 \end{pmatrix} \tag{3.3}$$

where the operator

$$\tilde{A}_n \equiv \begin{bmatrix} P + P_n R P_n, & P_n r \\ T P_n \quad , & 0 \end{bmatrix}$$

will be nonsingular if and only if A_n is. Due to the duality of the projection method definition, both linear equations in (3.3) can be solved at the expense of one LU-factorization.

The following error estimate follows easily from Lemma 3.1.

Lemma 3.2. There exists a constant c such that for all x in some neighborhood $\Omega \subset D$ of x_0

$$\| V(x) - V_n(x) \| + \| g(x) - g_n(x) \| \leq c \| (I - P_n) V(x) \|$$

and

$$\| u_n^*(x) - u_n^*(x) \| + \| g'(x) - g_n'(x) \| \to 0 \quad \text{as } n \to \infty.$$

Now we assume that a suitable discretization $f_n(x)$ of $f(x)$ is given, so that, for example,

$$\| f_n(x_0) \| \leq c \| (I - \tilde{P}_n) x_0 \|$$

for some projection $\tilde{P}_n$ on X, and also so that

$$\| f_n'(x_0) - f'(x_0) \| \to 0 \quad \text{as } n \to \infty.$$

Then the following theorem estimating the error in approximating x_0, the solution to (2.7), by x_n, the solution to the system

$$\begin{aligned} f_n(x) &= 0 \\ g_n(x) &= 0 \end{aligned} \tag{3.4}$$

is a straightforward application of Theorem 3.3 in [9].

Theorem 3.3. There exists a neighborhood of x_0 so that for all n sufficiently large, a solution x_n to (3.4) exists and is unique in Ω. Moreover, there exists a constant c so that

$$\| x_n - x_0 \| \leq c(\| (I - \tilde{P}_n) x_0 \| + \| (I - P_n) V_0 \|) . \tag{3.5}$$

In the case that our problem is to compute a generalized turning point for $f(z,\lambda) : X = Z \times R \to PZ$, then we define

$$f_n(x) = P_n f(x) + (\hat{P} - \hat{P}_n)x \tag{3.6}$$

where for $(z,\lambda) \in Z \times R$, $\hat{P}(z,\lambda) = (Pz,\lambda)$ and $\hat{P}_n(z,\lambda) = (P_n z,\lambda)$. Then (3.5) becomes

$$\| z_n - z_0 \| + |\lambda_n - \lambda_0| \le k(\|(I - P_n)z_0\| + \| (I - P_n)v_0 \|) .$$

Estimates of the form (3.5) have already been obtained in [5] and [9].

We next consider in more detail the convergence of the critical parameter, i.e., $\lambda_n - \lambda_0$, in the case $X = Z \times R$. Expanding $f(x_n)$ about x_0 we obtain

$$f(x_n) = f(x_0) + f_z(x_0)(z_n - z_0) + f_\lambda(x_0)(\lambda_n - \lambda_0) + O(\| x_n - x_0 \|^2). \tag{3.7}$$

From (3.7) it follows that

$$\lambda_n - \lambda_0 = u_0^* f(x_n)/u_0^* f_\lambda(x_0) + O(\| x_n - x_0 \|^2). \tag{3.8}$$

In applications it may be that $u_0^* f(x_n)$ is substantially smaller than the residual, $f(x_n)$, itself. If, for example, $f_n(x)$ is given by (3.6), then $u_0^* f(x_n) = u_0^*(f(x_n) - P_n f(x_n)) = (I - P_n^*)u_0^* f(x_n)$. Thus we find the inequality

$$|\lambda_n - \lambda_0| \le c(\| (I - P_n^*)u_0^* \| \cdot \| f(x_n) \| + \| x_n - x_0 \|^2). \tag{3.9}$$

From (3.9) we see that the condition $\| (I - P_n^*)u_0^* \| \to 0$ as $n \to \infty$ would imply superconvergence. That this case can occur in the case of the Rayleigh-Ritz method and simple turning points is given in [5]. For self-adjoint problems and using the energy norm, the term $\| (I - P_n^*)u_0^* \|$ will be of the same order as the residual.

As a second application of (3.5) and (3.9), we consider the bifurcation case. As mentioned in the introduction, the solutions to $f_n(z,\lambda_0) = 0$ will not in general consist of intersecting curves while $f_n(z,\lambda_n) = 0$ will at the point (z_n,λ_n) where (z_n,λ_n) solves (3.4). Thus one can use the solutions to $f_n(z,\lambda_n) = 0$ to produce an approximating solution set that has the desired bifurcation structure. The estimate for $\lambda_n - \lambda_0$ in (3.8) can be used to develop error estimates to the order of the discretization error for these curves. This procedure is related to a method suggested by Weber [21].

4. Application to Boundary Value Problems.

Consider the problem

$$y'' + f(\lambda,t,y) = 0, \quad 0 < t < 1, \tag{4.1}$$

subject to the boundary conditions

$$y(0) = y(1) = 0 \ . \tag{4.2}$$

With the substitution $z = y''$, (4.1) - (4.2) may be rewritten as

$$z + f(\lambda,t,Gz) = 0, \qquad z \in Z \equiv C[0,1] \ , \tag{4.3}$$

where G is the inverse operator for D^2 subject to (4.2). In the case of a simple turning point, the system (2.1) becomes

$$\begin{aligned} &(a) \quad v + f_y(\lambda,t,Gz)v - gr = 0, \qquad r \in Z \\ &(b) \quad Tv = 1 \ . \end{aligned} \tag{4.4}$$

Equation (4.4)(b) is a normalization condition. An example would be $Tv = (DGv)(0) = 1$. It is straightforward now to use Theorem 3.3 together with standard projection methods for (4.3) - (4.4) and deduce convergence results. See, for example, the survey paper [22] for many possible choices for P_n. We focus here briefly on the phenomenon of superconvergence for the critical parameter. We let P_n generate the method of collocation at Gauss points. For the case of the Ritz-Galerkin method and simple turning points, see [5]. In the case of collocation at Gauss points [4], the basic error estimates given in (3.8) - (3.9) do not give superconvergence directly because they are formulated in the "second-derivative" spaces. Although it is possible to establish higher-order rates using (3.8) - (3.9), it is easier to reformulate the systems (2.1) and (2.7) as boundary value problems and obtain the results directly. This reformulation is an extension of the technique given in [20].

For case of notation, we consider the simple turning point case. Then the system (2.1) defining g and the determining system (2.7) may be written together resulting in

$$\begin{aligned} &y'' + f(\lambda,t,y) = 0 && v'(0) = 1 \\ &y(0) = y(1) = 0 && \lambda' = 0 \\ &v'' + f_y(\lambda,t,y)v - gr = 0 && g' = 0 \\ &v(0) = v(1) = 0 && g(0) = 0 \end{aligned} \tag{4.5}$$

where the next to the last two equations in (4.5) force λ and g to be constants and the last equation then is equivalent to setting g to zero. The system (4.5) is simply a well-posed boundary value problem, and the system (3.4) represents its discretization using, in this case, collocation at Gauss points. Since λ_n will achieve superconvergence at the knots, then it will achieve superconvergence since it will be a constant. Thus, for example, if the basic polynomials are C^1-quartics, $|\lambda_n - \lambda_0| = O(h^6)$ would be obtained for sufficiently smooth functions f [1]. The same result would obtain for the bifurcation problem in the unfolding or perturbation parameter [16].

For the problem

$$y'' + 3.9 \exp[1/(1+\lambda y)] = 0, \qquad 0 < x < 1,$$

$$y(0) = y(1) = 0\ ,$$

we discretized using quartic splines and collocation at Gauss points and obtained the results in the following table:

h	$\lambda_h - \lambda_0$	Estimated order of convergence
1/4	.16 - 4	-
1/5	.38 - 5	6.3
1/6	.13 - 5	5.92
1/7	.52 - 6	5.98

Using central differences on the problem $y'' + \lambda e^y = 0$, $0 < x < 1$, and $y(0) = y(1) = 0$, there will be no superconvergence, and, indeed, we obtained the $O(h^2)$ results in the next table:

h	$\lambda_h - \lambda_0$
1/10	.0187
1/20	.0045
1/40	.0011

References

[1] J.P. Abbott, An efficient algorithm for the determination of certain bifurcation points, J. Comp. Appl. Math. 4 (1978), 19-27.

[2] U. Ascher, J. Christiansen and R.D. Russell, A collocation solver for mixed order systems of boundary value problems, Math. Comp. 33 (1979), 659-679.

[3] W.-J. Beyn and E. Bohl, Organizing centers for discrete reaction diffusion models, in "Numerical Methods for Bifurcation Problems," ed. T. Küpper et.al., Birkhauser (1984), 57-78.

[4] C. deBoor and B.K. Swartz, Collocation at Gaussian points, SIAM J. Numer. Anal. 10 (1973), 582-606.

[5] F. Brezzi, T. Rappaz and P. Raviart, Finite dimensional approximation of nonlinear problems. Part I. Branches of non-singular solutions, Numer. Math. 36 (1980), 1-25; Part II. Limit points, Numer. Math. 37 (1981), 1-28; Part III. Simple bifurcation points, Numer. Math. 38 (1981), 1-30.

[6] M.G. Crandall and P.H. Rabinowitz, Bifurcation from simple eigenvalues, J. Functional Analysis 9 (1979), 321-340.

[7] D. Dellwo, H.B. Keller, B.J. Matkowsky and E.L. Reiss, On the birth of isolas, SIAM J. Applied Math. 42 (1980), 956-963.

[8] E.J. Doedel, Auto: A program for the automatic bifurcation analysis of autonomous systems, Congressus Numerantium 30 (1981), 265-284.

[9] J.P. Fink and W. C. Rheinboldt, On the discretization error of parameterized nonlinear equations, Univ. of Pittsburgh, Inst. for Comp. Math. and Applications, Tech. Report ICMA-83-59, June, 1983.

[10] C.G. Gibson, Singular points of smooth mappings, Research Notes in Mathematics 25, Pitman, 1979.

[11] M. Golubitsky and B. L. Keyfitz, A qualitative study of the steady state solutions for a continuous flow stirred chemical reactor, SIAM J. Math. Anal. II (1980), 316-339.

[12] A. Griewank and G.W. Reddien, Characterization and computation of generalized turning points, SIAM J. Numer. Anal. 21 (1984), 176-185.

[13] A. Griewank and G.W. Reddien, Computation of turning and bifurcation points for two-point boundary value problems, in Proceedings of Seventh Annual Lecture Series in the Mathematical Sciences, Univ. of Arkansas, Fayetteville, 1983.

[14] H.B. Keller, Numerical solution of bifurcation and nonlinear eigenvalue problems, in Applications of Bifurcation Theory, ed. P. Rabinowitz, Academic Press, New York (1977), 359-384.

[15] T. Kupper, H.D. Mittelmann and H. Weber, Numerical Methods for Bifurcation Problems, ISNM 70, Birkhauser, Boston, 1984.

[16] H.D. Mittelmann and H. Weber, Bifurcation Problems and Their Numerical Solution, ISNM 54, Birkhauser, Boston, 1980.

[17] G. Moore and A. Spence, The calculation of turning points of nonlinear equations, SIAM J. Numer. Anal. 17 (1980), 567-576.

[18] G. Pönisch and M. Schwetlick, Computing turning points of curves implicitly defined by nonlinear equations depending on a parameter, Computing 26 (1981), 107-121.

[19] G.W. Reddien, Projection methods for two-point boundary value problems, SIAM Review 22 (1980), 156-171.

[20] R. Seydel, Numerical computation of branch points in ordinary differential equations, Numer. Math. 32 (1979), 51-68.

[21] H. Weber, On the numerical approximation of secondary bifurcation problems, in Numerical Solution of Nonlinear Equations, Lecture Notes in Math. 878, Springer-Verlag, Berlin (1981), 407-426.

[22] B. Werner and A. Spence, The computation of symmetry-breaking bifurcation points, SIAM J. Numer. Anal. 21 (1984), 388-399.

Department of Mathematics
Southern Methodist University
Dallas, Texas 75275

Progress in Scientific Computing, Vol. 5
Numerical Boundary Value ODEs

CALCULATING THE LOSS OF STABILITY BY TRANSIENT METHODS, WITH APPLICATION TO PARABOLIC PARTIAL DIFFERENTIAL EQUATIONS

R. Seydel

The calculation of Hopf bifurcations in systems of parabolic PDEs (one space variable) is considered. By semi-discretization via method of lines one obtains ODE systems, thereby enabling the usage of ODE methods. Some novel results are presented dealing with transient methods, i.e. methods that handle the steady state as a special periodic solution with amplidude zero. The results include the calculation of stability of periodic orbits as well as the computation of points of loss of stability.

1. The problem

Many problems in applied sciences as for instance reaction-diffusion problems are described by parabolic partial differential equations (PDE). In the case of one space variable (x) such a system may be given by ℓ equations $(i=1,2,..,\ell)$,

$$\begin{aligned}
\frac{\partial y_i}{\partial t} &= d_i \frac{\partial^2 y_i}{\partial x^2} + c_i \frac{\partial y_i}{\partial x} + f_i(y_1,\dots,y_n,\lambda) \\
\alpha_i^0 \frac{\partial y_i(0,t)}{\partial x} &+ \beta_i^0\, y_i(0,t) = \gamma_i^0 \qquad (1)\\
\alpha_i^1 \frac{\partial y_i(1,t)}{\partial x} &+ \beta_i^1\, y_i(1,t) = \gamma_i^1
\end{aligned}$$

Solutions $y(x,t)$ of (1) depend on the real parameter λ and on the real coefficients d_i, c_i, α_i, .. .

Steady-state solutions are defined by $\partial y/\partial t=0$, resulting in a boundary-value problem of ODE. The steady-state solutions may be stable only for certain values of λ, say for $\lambda<\lambda_0$. Under certain generic conditions the critical value λ_0 is a point of Hopf bifurcation [4], that is to say, there are periodic solutions in each neighborhood.

System (1) can be semidiscretized. Using the method of lines and, for simplicity, an equidistant grid with N interior points one has

$$h:=\frac{1}{N+1}\ ,\quad x_k:=k\cdot h \quad \text{for } k=0,1,\ldots,N,N+1\ .$$

Now, the system for $y_i(t,x_k)$ can be written as a first order system of ODE. With $y_{ik}(t):=y_i(t,x_k)$ one obtains a system of $n=N\cdot\ell$ equations,

$$\frac{\partial y_{ik}}{\partial t}=g_{ik}(y_{11},\ldots,y_{\ell N},\lambda) \tag{2}$$

which can be rewritten as

$$\dot{z}=g(z,\lambda)\ . \tag{2'}$$

The global numerical problem is to detect a Hopf bifurcation of (1) during tracing a steady-state branch, and to calculate periodic profiles.

2. Known methods; general loss of stability

All methods developed for ODE can be applied via (2). So-called indirect methods calculate the eigenvalues of the linearizations of available steady-state solutions; Hopf bifurcation is indicated by a loss of stability. Recently, "direct" methods for the calculation of critical solutions found some interest in literature; see for instance [9]. In the context of Hopf bifurcation there are essentially two kinds of direct methods, they may be denoted by "steady -state" and "transient" methods.

1. Steady-state methods solve suitably enlarged steady-state equations.
   a) ODE methods are proposed in [5, 6, 3]. Here, one considers the steady-state situation of the ODE system (2) which is described by the "algebraic" equations $g(z,\lambda)=0$.
   b) PDE methods [8, 10] are analoga to ODE approaches. Here, the steady-state systems corresponding to (1) are solved, which include ODE boundary-value problems.
2. Transient methods handle the Hopf point as a special periodic solution with amplitude zero. Such a method was proposed in [12] for the ODE case. This method solves the boundary-value problem (3).

$$\begin{pmatrix} z \\ T \\ \lambda \\ v \end{pmatrix}' = \begin{pmatrix} T\cdot g(z,\lambda) \\ 0 \\ 0 \\ T\cdot g_z(z,\lambda)\cdot v \end{pmatrix} , \quad \begin{pmatrix} z(0)-z(1) \\ v_1'(0) \\ v_1(0)-1 \\ v(0)-v(1) \end{pmatrix} = 0 \qquad (3)$$

In (3), ' denotes the differentiation with respect to the normalized time, T is the period, $v(t)$ a n-vector.

We first show that the equation (3) results as a special case from the general method [11] for calculating branch points. To see this, consider (z,T) as a vector function with $n^*=n+1$ components. The problem of calculating periodic solutions can be transformed to the boundary-value problem

$$\begin{pmatrix} z \\ T \end{pmatrix}' = \begin{pmatrix} T\cdot g(z,\lambda) \\ 0 \end{pmatrix} , \quad \begin{pmatrix} z(0)-z(1) \\ g_1(z(0),\lambda) \end{pmatrix} = 0 \quad .$$

Now, upon attaching this boundary-value problem to that one of its linearization

$$\begin{pmatrix} v \\ \theta \end{pmatrix}' = \begin{pmatrix} T\cdot g_z(z,\lambda)\cdot v + \theta\cdot g(z,\lambda) \\ 0 \end{pmatrix} , \quad \begin{pmatrix} v(0)-v(1) \\ \sum_j v_j \, \partial g_1/\partial z_j \end{pmatrix} = 0$$

together with $\lambda'=0$, $v_1(0)=1$ yields the branching system (3') for periodic solutions of dimension 2n*+1=2n+3,

$$\begin{pmatrix} z \\ T \\ \lambda \\ v \\ \theta \end{pmatrix}' = \begin{pmatrix} T\cdot g(z,\lambda) \\ 0 \\ 0 \\ T\cdot g_z(z,\lambda)\cdot v + \theta\cdot g(z,\lambda) \\ 0 \end{pmatrix} \qquad \begin{pmatrix} z(0)-z(1) \\ g_1(z(0),\lambda) \\ v(0)-v(1) \\ \sum_j v_j\,\partial g_1/\partial z_j \\ v_1(0)-1 \end{pmatrix} = 0 \qquad (3')$$

In the Hopf bifurcation case one has $g(z,\lambda)=0$ and thus system (3') can be reduced to the (2n+2)-system in (3); system (3) is included in the present version of the bifurcation software BIFPAC [13]. As will be shown later on, (3) can be reduced even further. Note that the full branching system (3') can be used to calculate the more general loss of stability of periodic solutions; in the following we merely focus on the special case of Hopf bifurcation. Some of the aspects to be outlined in the sequel hold true in the general case.

A major disadvantage of both transient method and ODE steady-state methods in the PDE context is the discretization in two variables, x and t. A PDE steady-state method needs only a discretization in x. In the present paper we are going to discuss ways how to establish a transient method efficient enough to meet the demands of large ODE systems as they may arise in the PDE setting.

3. Calculation of stability

Before discussing the loss of stability in some detail, we have to address ourselves to the essential tool of the calculation of the stability. Recently, interest has focussed on possible routes to turbulence [2]; corresponding parameter studies require repeated determination of stability. In this section, we propose a convenient algorithm for obtaining the monodromy matrix, whose eigenvalues determine stability.

First, we shortly recall definition and application of the monodromy matrix. Let $z(t)$ be a periodic solution of

(2) with period T that belongs to a particular parameter value λ. Let $A(t)$ be solution of the matrix differential equation

$$\dot{A} = \frac{\partial g(z,\lambda)}{\partial z} A \quad , \quad A(0) = I \tag{4}$$

In this initial-value problem I denotes the identity matrix.

Definition: $M = M(\lambda) := A(T)$ is called the monodromy matrix. For periodic z one eigenvalue of M is always unity. The other $n-1$ eigenvalues of M determine the stability and bifurcation behavior of periodic orbits [1]. Let us summarize briefly ($\bigcirc$ means unit circle in the complex plane):

- stable if all eigenvalues are inside $\bigcirc$
- unstable if there is an eigenvalue outside $\bigcirc$
- period doubling if one eigenvalue crosses $\bigcirc$ at -1
- birth or annihilation of limit cycle, transcritical or pitchfork bifurcation if one eigenvalue crosses $\bigcirc$ at $+1$
- bifurcation into tori if a pair of eigenvalues crosses $\bigcirc$ with nonzero imaginary part (resonance has to be excluded)

Let z_0 be a vector of initial values on the profile of $z(t)$. The definition of M suggests the integration of the initial-value problem

$$\begin{pmatrix} \dot{z} \\ \dot{A} \end{pmatrix} = \begin{pmatrix} g(z,\lambda) \\ g_z(z,\lambda)\cdot A \end{pmatrix} \quad , \quad \begin{pmatrix} z(0)-z_0 \\ A(0)-I \end{pmatrix} = 0 \quad . \tag{5}$$

By integrating (5) until $t=T$ one obtains the monodromy matrix. This one system of equation (5) is convenient for small or moderate values of n. In such a situation the dimension $n+n^2$ of (5) may be tolerable.

The dimension and storage requirements can be significantly reduced if system (3) is used. To this end, let us denote by e_k the k-th unit vector of $\mathbb{R}^n$, and let a_k be the k-th column of M. It can be easily seen that the following algorithm calculates the monodromy matrix.

Algorithm for calculating the monodromy matrix.

for $j=1,\ldots,n$ do:

integrate (3) for $0 \leq t \leq 1$ with initial values
$z(0)=z_0$, $v(0)=e_j$
Then $a_j=v(1)$.

Upon carrying out this algorithm one integrates n times a $(2n+2)$-system. Thereby, storage problems are by far not as stringent as in integrating (5). Note that this algorithm takes about twice as long as integrating (4). Of course, the two trivial differential equations for λ and T could be dropped.

It is interesting to realize that the monodromy matrix needs not to be computed if the periodic solution has been calculated by multiple shooting code in condensed form (see e.g. [14]); in this case M is immediately available. This can be seen as follows. Let E be the iteration matrix of multiple shooting ($E=A+BG_{m-1}\cdots G_1$ in the notation of [14], p.486). Then the n^2 matrix equation

$$M = I - E \qquad (6)$$

holds. The rank-deficiency 1 of the n^2 iteration matrix E due to the eigenvalue 1 of M does not affect the convergence of shooting if one embeds (2) into the $(n+2)$-subsystem of (3), see [12]. Then the full $(n+2)^2$-iteration matrix is usually nonsingular. One just takes the n^2 submatrix and obtains M via (6). Note that the n^2 matrix E has rank-deficiency 2 at bifurcation points of the type where an eigenvalue crosses $\bigcirc$ at +1 .

Let us note the two recent papers [15, 16] that base their numerical stability analysis on the eigenvalues of the monodromy matrix. The standard approach for calculating the monodromy matrix is (5); in [16] and apparently also in [15] this calculation can be part of the procedure for solving boundary-value problems. The two procedures proposed on this page are new.

4. A simple method for obtaining an initial guess

Before solving an enlarged system like (3) one needs some good initial guess. Values for z, λ and T are obtained very easily in tracing the steady-state branch. Yet a reasonable initial guess for v is still needed. In this section, we present a simple device.

The normalization of v in (3),

$$v_k(0)=1 \ , \quad v_k'(0)=0 \quad \text{with e.g. } k=1 \ , \tag{7}$$

is such that the k-th v-component of a solution of (3) is a priori known,

$$v_k(t) = \cos 2\pi t \quad . \tag{8}$$

Without loss of generality, the index k in the equations (7) and (8) can be taken to be $k=1$. Of course, one takes (8) as initial guess for the k-th component of v. What about the other $n-1$ components? By substituting (8) into the differential equation of (3) one has

$$v_k' = T \cdot \sum_{j=1}^{n} \frac{\partial g_k(z,\lambda)}{\partial z_j} v_j \qquad (g_k \text{ here k-th component of } g)$$

This results in one equation for $n-1$ unknowns,

$$\sum_{j \neq k} \frac{\partial g_k}{\partial z_j} v_j(t) = - \frac{2\pi}{T} \sin 2\pi t - \frac{\partial g_k}{\partial z_k} \cos 2\pi t \quad . \tag{9}$$

In case $n=2$, the two equations (8), (9) define an initial guess for v which is both cheap and precise. In applications (1) often consists of $\ell=2$ equations. Thus the case $n=2$ frequently arises when one uses one line only ($N=1$). In case $\ell>2$ ($n>2$), by (9) at least a crude guess for v is obtainable, good enough in many applications. Allowing more expense, an initial guess could be obtained using eigenvectors. Note that $v(0)$ is eigenvector of the monodromy matrix belonging to the double eigenvalue 1. For larger N, an initial guess is obtained based on the lower grids.

5. A modification of the transient method

Upon implementing the PDE method [8] of solving a system of three boundary-value problems of second order, one can formulate this method as one first-order system of $6\ell+2$ equations; this includes two trivial equations for the unknown values of parameter and period. This number $6\ell+2$ should be a sort of "limit" if one developes a transient method based on the semi-discretization (2).

This limiting level implies the bound $N \leqslant 3$ on the admitted number of interior lines, if (3) is used for the calculation of the Hopf point. However, the approach (3) can be modified to solving a mixed system: the first n components of (3) can be replaced by the steady-state equations. This novel method for the calculation of Hopf points will be described in detail in a forthcoming paper; we just give a main idea: The linearization of (1) is

$$\frac{\partial h_i}{\partial t} = d_i\, h_i'' + c_i\, h_i' + \sum_{j=1}^{\ell} \frac{\partial f_i}{\partial y_j}\, h_j \qquad \begin{array}{l}(h' \text{ for } \partial h/\partial x)\\ + \text{ boundary}\\ \quad\text{conditions}\end{array}$$

Now, only this linearization is semidiscretized, yielding $N\ell$ equations (10),

$$\dot{h}_{ik} = T\cdot\tilde{g}_{ik}(\ldots) \qquad , \qquad h_{ik}(0)-h_{ik}(1)=0 \tag{10}$$

In (10), t-dependent periodic solutions are to be calculated. Upon attaching the x-dependent steady-state system together with the trivial differential equations

$$\lambda' = T' = 0$$

yields a system of $\ell N+2\ell+2$ first order differential equations. Hereby, the system size of $6\ell+2$ is achieved with the number of lines of $N=4$. The gain in efficiency is significant if the number of equations is high. Thus, this new approach may be valuable in our context of semidiscretizing a system of partial differential equations.

6. Concluding Remarks

Theoretical considerations indicate that system (3) might show poor convergence properties due to an underlying singularity. However, this is not the case. Extensive use of (3) lead to results that clearly reveal good convergence. Typically, the rate of convergence is such that after an amount of work equivalent to 2-3 Newton steps the critical parameter is approximated to 6 digits (CYBER with 48-bit mantissa). The speed of convergence compares favorably to the average time required to calculate a periodic oscillation.

As noted in Section 3, the algorithm that calculates a monodromy matrix by integrating system (3) n times is about twice as slow as the standard approach (5). This disadvantage can be removed if the $z(t)$ of the first n components is accurately approximated, say by a Hermite polynomial. The gain in efficiency, however, is compensated by the need of a special routine with complicated error control.

The method briefly described in the foregoing section is implemented in a (prototype)routine; the preliminary numerical results are satisfactory. Note that if one needs only a low-accuracy guess of the Hopf point (using up to three lines), the transient method proposed in this paper needs less storage than the PDE method [8] . The code based on the present method improves the results of several lines by extrapolation. Thereby the first crude guess (based on only one line) is provided by (3).

References

1 V.I. Arnold: Geometrical methods in the Theory of Ordinary Differential Equations. Springer, New York, 1983

2 J.-P. Eckmann: Roads to turbulence in dissipative dynamical systems. Reviews of Modern Physics 53 (1981) 643-654

3 A. Griewank, G.W. Reddien: The calculation of Hopf points by a direct method. IMA J. Numer. Anal. 3 (1983) 295-303

4 B.D. Hassard, N.D. Kazarinoff, Y.H. Wan: Theory and applications of Hopf bifurcations. Cambridge University Press, Cambridge 1981

5 M. Holodniok, M. Kubíček: New algorithms for evaluation of complex bifurcation points in ordinary differential equations. A comparative numerical study. Manuscript 1981, to be published

6 A.D. Jepson: Numerical Hopf bifurcation. Thesis, California Institute of Technology, Pasadena, 1981

7 H.B. Keller: Numerical solution of bifurcation and nonlinear eigenvalue problems. in: P.H. Rabinowitz: Applications of bifurcation theory. New York, Academic Press 1977

8 M. Kubíček, M. Holodniok: Evaluating of Hopf bifurcation points in parabolic equations describing heat and mass transfer in chemical reactors. CES 39 (1984) 593-599

9 R.G. Melhem, W.C. Rheinboldt: A comparison of methods for determining turning points of nonlinear equations. Computing 29 (1982) 201-226

10 D. Roose: An algorithm for the computation of Hopf bifurcation points. Manuscript, Leuven 1984

11 R. Seydel: Numerische Berechnung von Verzweigungen bei gewöhnlichen Differentialgleichungen. Report 7736, TU Munich, 1977 (see Numer. Math. 32 (1979) 51-68)

12 R. Seydel: Numerical computation of periodic orbits that bifurcate from stationary solutions of ordinary differential equations. Appl. Math. Comp. 9 (1981) 257-271

13 R. Seydel: BIFPAC - a program package for calculating bifurcations. Buffalo 1983

14 J. Stoer, R. Bulirsch: Introduction to numerical analysis. Springer, New York, 1980

15 E.J. Doedel, R.F. Heinemann: Numerical Computation of periodic solution branches and oscillatory dynamics of the stirred tank reactor with A→B→C reactions. Chemical Engineering Science 38 (1983) 1493-1499

16 M. Holodniok, M. Kubíček: Derper - An algorithm for continuation of periodic solutions in ordinary differential equations. J. Comput. Physics, in press

R. Seydel
Department of Mathematics
University at Buffalo , N.Y. 14214

Progress in Scientific Computing, Vol. 5
Numerical Boundary Value ODEs

A RUNGE-KUTTA-NYSTROM METHOD FOR DELAY DIFFERENTIAL EQUATIONS*

By A. Bellen.

1. INTRODUCTION

Let consider the following boundary value problem for second order delay differential systems:

$$
\begin{aligned}
& y''(t)=f(t,y(t),y'(t),y(t-\tau(t)),y'(t-\sigma(t))) \quad t_\circ \leq t \leq b \\
& y(t)=\phi(t) \qquad t \leq t_\circ \\
& y'(t)=\phi'(t) \qquad t < t_\circ \\
& y(b)=y_b
\end{aligned}
\tag{1}
$$

$y: \mathbb{R} \to \mathbb{R}^m$, $f: [t_\circ,b] \times \mathbb{R}^{4m} \to \mathbb{R}$ and $\tau(t), \sigma(t) > 0$.

One easily realizes that, in general, at the point $t_\circ$ the solution does not join smoothly the initial function ϕ, and therefore jump discontinuities in y'' can occur at any point $\bar{t} \in (t_\circ,b)$ such that $\sigma(\bar{t})=t_\circ$. Similarly y''' is discontinuous at any point $\bar{\bar{t}} \in (t_\circ,b)$ such that either $\tau(\bar{\bar{t}})=t_\circ$ or $\sigma(\bar{\bar{t}})=\bar{t}$, and so on by increasing the order of derivatives. Remark that, for the positivity of the delays, $\bar{\bar{t}}>\bar{t}$ and jump discontinuities in the higher order derivatives shift forward as the order increases, so that after a finite number of differentiations such discontinuity points go outside the interval $(t_\circ,b)$. Other jump discontinuities in the higher order derivatives can rise from discontinuities in f (with

* Work performed within the activity of C.N.R. (Italian National Council of Research) Prog. Final. "Informatica" Sottoprog. P1 - SOFMAT.

respect to t), τ, σ or ϕ. We shall call "breaking points" all these discontinuity points ξ_i. Remark that between consecutive breaking points the function y" is as smooth as $f(t,\cdot,\cdot,\cdot,\cdot)$.

So far, very few are the papers on numerical solutions of two sides b.v.p. for DDE's (Delay differential equations). Among them we can quote: i) methods which reduce the DDE into a larger system of ODE's [1], ii) iterative methods [3], [7], [9], iii) global methods [2], [5], [12].

Here we treat the problem by means of collocation in piecewise polynomial spaces either in the shooting or in the global approach. In both cases, under some conditions on the delays and on the choice of the mesh, the superconvergence phenomenon, well-known for ODE's, extends to DDE's.

Alternatively, in the shooting approach, an iterated-defect-correction like method is presented which allows to treat some more general cases and to drop the conditions on the choice of the mesh.

2. THE SHOOTING APPROACH.

In solving numerically the equation (1) by means of shooting, one should have a code available for the following second order delay initial value problem:

$$\begin{aligned} &y''(t)=f(t,y(t),y'(t),y(t-\tau(t)),y'(t-\sigma(t))) \quad t_o\leq t\leq b \\ &y(t)=\phi(t) \quad \text{and} \quad y'(t)=\phi'(t) \qquad t<t_o \\ &y(t_o)=\phi(t_o) \\ &y'(t_o)=y'_o. \end{aligned} \tag{2}$$

To this purpose, we consider a method based on the one-step collocation at n shifted gaussian points on each subinterval of the mesh Δ, with approximate solution u_Δ in the class $S^1_{\Delta,n+1}$ of C^1 piecewise polynomials of degree n+1 on each sub interval of Δ. The method, which is equivalent to an n-level implicit Runge-Kutta-Nystrom method, is well-known for an ordinary i.v.p.: $z''=g(t,z,z')$, $z(t_\circ)=z_\circ$, $z'(t_\circ)=z'_\circ$, for which the most direct and well-turned proof of the order results is based on the Alekseev-Gröbner formula and was given by Nørsett-Wanner [11] and Nørsett [10]. The results are:

if $z\in C^{2n+2}(t_\circ,b)$, then

$$\max_{t_k\in\Delta} |u_\Delta^{(j)}(t_k)-z^{(j)}(t_k)|=O(|\Delta|^{2n}) \qquad j=0,1 \tag{2'}$$

$$\max_{t_k\in\Delta} \max_{t\in(t_k,t_{k+1})} |u_\Delta^{(j)}(t)-z^{(j)}(t)|=O(|\Delta|^{n+2-j}) \qquad j=0,1,\dots,n+1 \tag{2''}$$

where $|\Delta|=\max_{t_k\in\Delta}|t_{k+1}-t_k|$.

In dealing with delay i.v.p!s, the approximate solution u_Δ of (2) is inductively defined, between consecutive breaking points ξ_i,ξ_{i+1} , as the one-step collocation approximation of the ordinary i.v.p.: (let's use the abbreviation: $f_{u_\Delta}(t,z,z'):=f(t,z(t),z'(t),u_\Delta(t-\tau(t)),u_\Delta(t-\sigma(t)))$)

$$\begin{aligned} z''(t)&=f_{u_\Delta}(t,z,z') \qquad t\in[\xi_i,\xi_{i+1}] \\ z(\xi_i)&=u_\Delta(\xi_i) \\ z'(\xi_i)&=u'_\Delta(\xi_i) \end{aligned} \tag{3}$$

$i=0,1,\dots,s$; $\xi_\circ=t_\circ$; $\xi_{s+1}\geq b$; $u_\Delta(t)=\phi(t)$ and $u'_\Delta(t)=\phi'(t)$ for $t<t_\circ$; $u_\Delta(t_\circ)=\phi(t_\circ)$ and $u'_\Delta(t_\circ)=y'_\circ$.

Henceforward, the foregoing inductively defined method will be referred to as the block-by-block collocation method, the blocks I_i being the intervals $[\xi_{i-1},\xi_i]$ $i=1,\dots,s+1$.

Assume the solution y of (2) to be of class C^{2n+2} on each block. As in the first order case (see[4]),the expected result is that superconvergence rate 2n at the mesh points still occur despite the method make use of values of u_Δ and u'_Δ, at non nodal points, where the accuracy order is only n+2 and n+1 respectively. According to the first order delay i.v.p., the superconvergence results are achieved at the cost of a severe restriction in the choice of the mesh

Namely, a part the first interval I_1 where the nodes $t_0=t_0^1<$ $<t_1^1<\ldots<t_N^1=\xi_1$ are choosen arbitrarily, the delays must map each subinterval $[t_k^i,t_{k+1}^i]$ of I_i into the subinterval $[t_k^j,t_{k+1}^j]$ of I_j for some $j<i$. Such a condition reduces the feasibility of the method to special cases including $\tau(t)=\sigma(t)$, for variable delays, and $\sigma=n\tau$ (or $\tau=n\sigma$) $n\in\mathbb{N}$, for constant delays. Furthermore, assigned the mesh points in the first block I_1, the other mesh points $t_k^i\in I_i$ $i=2,\ldots,s+1$ must be computed by solving recursively $t_k^{i-1}=t_k^i-\tau(t_k^i)$ $k=1,\ldots$ $\ldots,N$.

Alternatively, one could fix the mesh points in the intervals $[\xi_s,b]$ and $[b-\tau(b),\xi_s]$ and then compute (easier than before) the rest of the mesh backward by the same recursion (which now turns out to be explicit). In this way one can also guarantee that the end point b belongs to the mesh, as required in the next chapter concerning with boundary value problems.

We shall call each of such meshes "constrained mesh".

Although the additional work for the precomputation of the mesh seems to make the method of theoretical interest rather than of practical utility, a recent stability analysis of numerical methods for DDE's (see [13]) shows that the constrained mesh selection produces, for certain methods including collocation, the largest absolute stability region.

It is interesting to remark that the restriction on the mesh means that the same mesh is used as would be in the transformed ODE's given by the method of Bellman (see [6]). Nevertheless, while for constant delays the two methods are exactly equivalent and provide the same numerical approximation, for variable delays Bellman's transformation requires the knowledge of the derivative of the functions $H(t):=(t-\tau(t))^{-1}$, $H^2(t):=H(H(t)),\ldots,H^{s+1}(t):=H(H^s(t))$. Moreover the j-th equation of the ODE's system is different from the equation rising in the j-th block of the block-by-block method (just by the presence of a term $d/dt(H^j(t))$).

So,for centain methods (collocation in this paper), Bellman's tranformation to ODE's and the block-by-block implementation give different results (though qualitatively similar).

We leave out proving the existence, for $|\Delta|$ small enough, and the uniform convergence result:

$$\max_{t_k^i\in\Delta} \max_{t\in(t_{k-1}^i,t_k^i)} |u_\Delta^{(j)}(t)-y^{(j)}(t)|=O(|\Delta|^{n+2-j}) \qquad j=0,\ldots,n+1$$

for the block-by-block collocation solution u_Δ.

Indeed the proofs can be carried out by standard arguments (see e.g. [4] for first order DDE's) and hold for any mesh including the breaking points.

Whereas we shall prove the following superconvergence theorem. The proof, which could be carried out by rearranging to second order equations the proof in [4], is performed in a quite different way. It draws inspiration from a perturbation analysis given in [15] where the author treats a first order initial value problem for DDE's of neutral type.

THEOREM 1. Let the delays $\tau(t)$ and $\sigma(t)$ fulfil the conditions described above and the mesh $\Delta=\{t_k^i\}$ $k=0,\ldots,N$ $i=1,\ldots,s+1$ be a constrained mesh. If the solution y of (2) is of class C^{2n+2} on each block I_i, then the approximate solution u_Δ, given by the block-by-block collocation method satisfies the superconvergence result:

$$\max_{t_k^i\in\Delta} |u_\Delta^{(j)}(t_k^i)-y^{(j)}(t_k^i)|=O(|\Delta|^{2n}) \qquad j=0,1 \tag{4}$$

Proof. Since in the first block the equation (3) reduces to an ODE, we have:

$$\max_{t_k^1\in\Delta} |u_\Delta^{(j)}(t_k^1)-y^{(j)}(t_k^1)|=O(|\Delta|^{2n}) \qquad j=0,1 \tag{5}$$

and

$$\max_{t_k^1\in\Delta}\ \max_{t\in(t_{k-1}^1,t_k^1)} |u_\Delta^{(j)}(t)-y^{(j)}(t)|=O(|\Delta|^{n+2-j}) \qquad j=0,1,\dots,n+1. \tag{6}$$

Moreover, $\forall G(t)\in C^{2n}[t_\circ,\xi_1]$, we have:

$$\int_{t_\circ}^{t_k^1} G(t)(u_\Delta''(t)-y''(t))dt=O(|\Delta|^{2n}) \qquad \forall k. \tag{7}$$

In fact, by the Alekseev-Gröbner formula,

$$u_\Delta'(t)-y'(t)=\int_{t_\circ}^{t}\Gamma(t,s)r(s)ds \qquad \forall t\in[t_\circ,\xi_1]$$

where $r(s)=u_\Delta''(s)-f_\phi(s,u_\Delta,u_\Delta')$, and hence:

$$u_\Delta''(t)-y''(t)=\int_{t_\circ}^{t}(\partial/\partial t)\Gamma(t,s)r(s)ds + r(t), \quad (\Gamma(t,t)=1)$$

$$\int_{t_\circ}^{\rho}G(t)(u_\Delta''(t)-y''(t))dt=\int_{t_\circ}^{\rho}\int_{t_\circ}^{t}G(t)(\partial/\partial t)\Gamma(t,s)r(s)dsdt+$$

$$+\int_{t_\circ}^{\rho}G(t)r(t)dt=\int_{t_\circ}^{\rho}H(\rho,s)r(s)ds \qquad \forall \rho\in[t_\circ,\xi_1]$$

where:

$$H(\rho,s)=G(s)+\int_{s}^{\rho}G(\xi)(\partial/\partial t)\Gamma(\xi,s)d\xi.$$

At any node t_k^1,

$$\int_{t_\circ}^{t_k^1}G(t)(u_\Delta''(t)-y''(t))dt=\int_{t_\circ}^{t_k^1}H(t_k^1,s)r(s)ds=O(|\Delta|^{2n})$$

since $r(s)$ vanishes at the collocation points (which are the nodes of the composite Gauss-Legendre quadrature rule), and its 2n derivatives are uniformly bounded as well as those of $H(t_k^1,s)$. So (7) is proved. Integration by parts of (7) yields:

$$\int_{t_\circ}^{t_k^1}G'(s)(u_\Delta'(s)-y'(s))ds=O(|\Delta|^{2n}) \tag{8}$$

and

$$\int_{t_\circ}^{t_k^1}G''(s)(u_\Delta(s)-y(s))ds=O(|\Delta|^{2n}) \tag{9}$$

Where G' and G" (belonging to C^{2n}) are arbitrary by the arbitrariness of G.

Now, by induction on the blocks, assume the properties (5-9) to hold for the nodes of the blocks $I_1,\ldots I_{i-1}$, and prove it for the nodes of I_i.

In order to prove that, $\forall G(t)\in C^{2n}[\xi_{i-1},\xi_i]$,

$$\int_{\xi_i}^{t_k^i} G(t)(u_\Delta''(t)-y''(t))dt=O(|\Delta|^{2n}), \qquad \forall k \tag{10}$$

let w(t) satisfy the ordinary i.v.p.

$$\begin{aligned} &w''(t)=f_y(t,w,w') \qquad \xi_{i-1}\le t\le\xi_i \\ &w(\xi_{i-1})=u_\Delta(\xi_{i-1}) \\ &w'(\xi_{i-1})=u_\Delta'(\xi_{i-1}). \end{aligned} \tag{11}$$

By the smoothness conditions on f, and the continuous dependence on the initial data, we have:

$$\max_{t\in I_i} |w^{(j)}(t)-y^{(j)}(t)|=O(|\Delta|^{2n}) \qquad j=0,1$$

and hence:

$$\max_{t\in I_i} |w''(t)-y''(t)|\le L\|w-y\|_{C^1}=O(|\Delta|^{2n}). \tag{12}$$

Therefore, by (12),

$$\int_{\xi_i}^{t_k^i} G(t)(u_\Delta''(t)-y''(t))dt=\int_{\xi_i}^{t_k^i} G(t)(u_\Delta''(t)-w''(t))dt+ +O(|\Delta|^{2n}) \quad \forall k. \tag{13}$$

The Alekseev-Gröbner formula for the problem (11)

$$u_\Delta'(t)-w'(t)=\int_{\xi_i}^{t} \Gamma(t,s)(u_\Delta''(s)-f_y(s,u_\Delta,u_\Delta'))ds,$$

yields, as above,

$$\int_{\xi_i}^{t_k^i} G(t)(u_\Delta''(t)-w''(t))dt=\int_{\xi_i}^{t_k^i} H(t_k^i,s)(u_\Delta''(s)-f_y(s,u_\Delta,u_\Delta'))ds.$$

By the Taylor expansion theorem, and the inductive hypothesis (6) till the block I_{i-1}, we get:

$$\int_{\xi_i}^{t_k^i} G(t)(u''(t)-w''(t))dt = \int_{\xi_i}^{t_k^i} H(t_k^i,s)(u''_\Delta(s)-f_{u_\Delta}(s,u_\Delta,u'_\Delta))ds +$$

$$+ \int_{\xi_i}^{t_k^i} H(t_k^i,s)(\partial/\partial y)f_y(s,u_\Delta,u'_\Delta)(u_\Delta(s-\tau(s))-y(s-\sigma(s)))ds +$$

$$+ \int_{\xi_i}^{t_k^i} H(t_k^i,s)(\partial/\partial y')f_y(s,u_\Delta,u'_\Delta)(u'_\Delta(s-\tau(s))-y'(s-\sigma(s)))ds +$$

$$+O(|\Delta|^{2n+2}).$$

By the gaussian quadrature rule, the first integral is an $O(|\Delta|^{2n})$, and by the inductive hypotheses (8)-(9) till I_{i-1}, together with the conditions on the nodes of I_i, the last two integrals are $O(|\Delta|^{2n})$ too. Thus, by (13), (10) is proved at the nodes of I_i.

The equality (10) for $G(t)=I$, yields, by the inductive hypothesis (5)(for $j=1$):

$$u'_\Delta(t_k^i)-y'(t_k^i)=u'_\Delta(\xi_{i-1})-y'(\xi_{i-1})+O(|\Delta|^{2n})=O(|\Delta|^{2n}) \quad (14)$$

and integration by parts of (10) yields:

$$\int_{\xi_i}^{t_k^i} G'(t)(u'_\Delta(t)-y'(t))dt=O(|\Delta|^{2n}) \quad (15)$$

for all k. By acting similarly on (15), we get:

$$\int_{\xi_i}^{t_k^i} G''(t)(u_\Delta(t)-y(t))dt=O(|\Delta|^{2n})$$

and

$$u_\Delta(t_k^i)-y(t_k^i)=O(|\Delta|^{2n}) \quad (16)$$

for all k. Finally (6) is easily proved by (14),(16) and the smoothness of f. So the properties (5-9) hold on I_i and the proof is complete.

In order to consider also the case $\tau(t)\neq\sigma(t)$ and to remove the constraints on the mesh preserving the superconvergence at the nodes, assume to have a method available to raise, block-by-block, the uniform accuracy order of the collocation approximation of both y and y' to 2n. Then define u* and u'* (not necessarily related by $u'^* = (\partial/\partial t)u^*$)

inductively on the blocks as the uniformly corrected approximations of order 2n of the collocation solution u_Δ of the ordinary i.v.p.:

$$\begin{aligned} &z''(t)=f_{u*}(t,z,z') \qquad t\in[\xi_i,\xi_{i+1}] \\ &z(\xi_i)=u^*(\xi_i) \\ &z'(\xi_i)=u'^*(\xi_i). \end{aligned} \tag{17}$$

It is not difficult to see that, if $\|y-u^*\|$ and $\|y'-u'^*\|$ in $[t_o,\xi_i]$ are both $O(|\Delta|^{2n})$, then, for any mesh in the block I_{i+1}, the collocation solution u_Δ of (17) satisfies, with respect to y, the order results (2')-(2"), since (17) can be regarded as a perturbation of magnitude $O(|\Delta|^{2n})$ of the problem: $z''(t)=f_y(t,z,z')$; $z(\xi_i)=y(\xi_i)$; $z'(\xi_i)=y'(\xi_i)$ the solution of which is y.

It remain to show how to get u^* and u'^* in I_{i+1}, by raising the accuracy of u_Δ and u'_Δ to 2n. To this purpose, let $\Pi_{\Delta,i}$ be the set of piecewise polynomials of degree i on each subinterval of Δ and, for any function g, $P_{\Delta,i}(g)$ be the element of $\Pi_{\Delta,i}$ interpolating g at i+1 points on each subinterval (for instance equidistant points).

Consider the following Iterated-Defect-Correction like method:

$$\begin{aligned} &w_o:=u_\Delta \quad (\in\Pi_{\Delta,n+1}), \\ &w_k:=w_{k-1}+\Phi_k \qquad k=1,2,\ldots \end{aligned} \tag{18}$$

where $\Phi_k\in\Pi_{\Delta,n+1+k}$ and is defined by the interpolatory conditions:

$$\Phi''_k=P_{\Delta,n+k-1}(f_{u*}(t,w_{k-1},w'_{k-1})-w''_{k-1}) \tag{19}$$

and

$$\lim_{t\to t_p^+}\Phi_k(t) = \lim_{t\to t_p^+}\Phi'_k(t) = 0 \tag{20}$$

at any node t_p of $\Delta\cap I_{i+1}$.

THEOREM 2. The sequence w_k fulfils the estimates:

$$\|y^{(j)}-w_k^{(j)}\|=O(|\Delta|^{\min(2n,n+k+2-j)}) \quad j=0,\dots,n+k+1. \tag{21}$$

Proof. Since $w_o=u_\Delta$, (21) holds for k=0; let's prove it for k=1.

$$y''(t)-w_1''(t)=f_y(t,y,y')-w_o''(t)-\Phi_1''(t) \tag{22}$$

and

$$\|y''-w_1''\| \leq \| f_y(t,y,y')-f_y(t,w_o,w_o')\|+\|f_y(t,w_o,w_o')-$$
$$-P_{\Delta,n}f_{u*}(t,w_o,w_o')\|$$

On the other hand:

$$\|f_y(t,w_o,w_o')-P_{\Delta,n}f_{u*}(t,w_o,w_o')\| \leq \|P_{\Delta,n}f_{u*}(t,w_o,w_o')-$$
$$P_{\Delta,n}f_y(t,w_o,w_o')\|+\|(I-P_{\Delta,n})f_y(t,w_o,w_o')\|$$

where, by the hypotheses on u* and u'* ,the first right hand term is an $O(|\Delta|^{2n})$ and, by the smoothness of f_y and the uniform boundedness of $w_o^{(j)}$ $j=0,\dots,n+1$, the second one is an $O(|\Delta|^{n+1})$. Therefore:

$$\|y''-w_1''\|=O(|\Delta|^{n+1}). \tag{23}$$

By differentiating (22), one easily gets (21) for k=1 and j>2.

For any $t\in I_{i+1}$, let t_p be the node of $\Delta\cap I_{i+1}$ such that $t_p\leq t\leq t_{p+1}$, then:

$$y'(t)-w_1'(t)=y'(t_p)-w_1'(t_p)+\int_{t_p}^{t}(y''(s)-w_1''(s))ds \tag{24}$$

where, by (20), $w_1'(t_p)=w_o'(t_p)$. Thus (23) yields:

$$\|y'-w_1'\| \leq O(|\Delta|^{2n})+O(|\Delta|^{n+2})=O(|\Delta|^{\min(2n,n+2)}).$$

Similarly, by integrating (24),

$$\| y-w_1\| \leq O(|\Delta|^{2n})+O(|\Delta|^{n+3})=O(|\Delta|^{\min(2n,n+3)})$$

so that (21) is proved for k=1.

By assuming (21) to be true for k=m, one easily proves it for m+1 by following the same scheme used for k=1. So the proof is complete.

For k=n-2, (21) yields $\|y-w_{n-2}\|=O(|\Delta|^{2n})$ and for k=n-1 yields $\|y'-w'_{n-1}\|=O(|\Delta|^{2n})$ on I_{i+1}. Therefore we can take $u^*=w_{n-2}$ and $u'^*=w'_{n-1}$ in the block I_{i+1} as the uniformly corrected approximations of order 2n, for y and y'.

In order to reduce the number of evaluations of f, one could take other projection operators instead of interpolation operators (see [14]).

3. THE GLOBAL APPROACH.

A global approach for getting an approximate solution of (1) is the well known collocation at gaussian points [8]:

$$u''_\Delta=P_{\Delta,n-1}f(t,u_\Delta(t),u'_\Delta(t),u_\Delta(t-\tau(t)),u'_\Delta(t-\sigma(t))) \quad (25)$$

$u_\Delta\in S^1_{\Delta,n+1}$ and interpolation points the zeroes of the shifted n-Legendre polynomials.

If $y_\circ$ is an isolated solution of (1) and the linearized equation at $y_\circ$ with vanishing boundary conditions has only the trivial solution, it can be proved by standard arguments (see [5] or [12]) that, for sufficiently small $|\Delta|$, (25) is uniquely solvable in a neighbourhood of $y_\circ$ and $\|u_\Delta\|_{C^2}\leq$const. In particular, u'_Δ is uniformly bounded. Therefore if z is the solution of the initial value problem:

$$\begin{aligned} &z''(t)=f_z(t,z,z') && t_\circ\leq t\leq b \\ &z(t)=\phi(t) && t\leq t_\circ \\ &z'(t)=\phi'(t) && t<t_\circ \\ &z'(t_\circ)=u'_\Delta(t_\circ) \end{aligned} \quad (26)$$

also the derivatives z' are uniformly bounded as $|\Delta|\to 0$.

On the other hand, the collocation solution of the i.v.p. (26) in the same space $S^1_{\Delta,n+1}$ and with the same collocation points, is just u_Δ. Thus, under the hypotheses of the theorem 1 on τ,σ and on the mesh Δ, we have:

$$\max_{t_k^i\in\Delta} |z^{(j)}(t_k^i)-u_\Delta^{(j)}(t_k^i)|=O(|\Delta|^{2n}) \qquad j=0,1.$$

In particular: $|z(b)-u_\Delta(b)|=|z(b)-y_b|=O(|\Delta|^{2n})$.

Since two solutions y^1 and y^2 of (1), relating to different end values y_b^1 and y_b^2, satisfy:

$$||y^1-y^2||_{C^2}\leq K|y_b^1-y_b^2|,$$

we get:

$$||y-z||_{C^2}=O(|\Delta|^{2n})$$

and hence:

$$\max_{t_k^i\in\Delta} |y^{(j)}(t_k^i)-u_\Delta^{(j)}(t_k^i)|=O(|\Delta|^{2n}) \qquad j=0,1.$$

To sum up, the superconvergence results of the collocation method for b.v.p.'s of ODE's extend to DDE's in the constrained mesh selection.

Also for b.v.p.'s we must remark that for constant delays the method above is equivalent to solving by collocation a transformed ODE system (see [1]), while for non constant delays the two methods are different.

For b.v.p.'s the iterated defect correction proposed for i.v.p.'s ought to be applied "a priori" and this increases the complicacy of the non linear system to be solved. So, in order to drop the constraints on the mesh, one must be satisfied with the lower order of convergence.

REFERENCES.

[1] Ascher U.,Russel R.D.: Reformulation of boundary value problems into "standard" form. SIAM Rew.23,238-254(1981).

[2] Bader G.: Solving boundary value problems for functional differential equations by collocation. These Proceed.

[3] Bellen A. Cohen's iteration process for boundary value problems for functional differential equations. Rend. Ist. Matem. Univ. Trieste II, 32-46 (1979).

[4] Bellen A.: One step collocation for delay differential equations. J.Comp.Appl.Math. 10, 275-283 (1984).

[5] Bellen A.,Zennaro M.: A collocation method for boundary value problems of differential equations with functional arguments. Computing 32, 307-318 (1984).

[6] Bellman R.,Cook K.L.: On the computational solution of a class of functional differential equations. J.Math. Anal.Appl. 12, 495-500 (1965).

[7] Chocholaty P.,Slahor L.:A method to boundary value problems for delay equations. Numer.Math. 33, 69-75 (1979).

[8] De Boor c.,Swartz B.: Collocation at gaussian points. SIAM J.Numer.Anal. 10, 582-606 (1973).

[9] Hutson V.C.L.: Boundary-value-problems for differential difference equations. J.Differential Equations 36, 363-373 (1980).

[10] Nørsett S.P.:Collocation and perturbed collocation. Lecture Notes in Math. 773 (Springer, Berlin 1980),119-132.

[11] Nørsett S.P.,Wanner G.: The real-pole sandwich for rational approximation and oscillation equations. BIT 19, 79-94 (1979).

[12] Reddien G.W.,Travis C.C.: Approximation methods for boundary value problems of differential equations with functional arguments. J.Math.Anal.Appl.46, 62-74 (1974).

[13] Zennaro M.: On the P-stability of one-step collocation for delay differential equations. To appear.

[14] Zennaro M.: One-step collocation: uniform superconvergence, predictor-corrector method, local error estimate. In press on SIAM J.Numer.Anal.

[15] Zennaro M.:Natural continuous extensions of Runge-Kutta methods. A preprint.

Progress in Scientific Computing, Vol. 5
Numerical Boundary Value ODEs

A Finite Difference Method for the Basic Stationary Semiconductor Device Equations

Peter A. Markowich

Institut für Angewandte und Numerische Mathematik,
TU Wien, Gußhausstraße 27-29,
A-1040 Wien, Austria

Abstract

In this paper we analyse a special-purpose finite difference scheme for the basic stationary semiconductor device equations in one space dimension. These equations model potential distribution, carrier concentration and current flow in an arbitrary one-dimensional semiconductor device and they consist of three second order ordinary differential equations subject to boundary conditions. A small parameter appears as multiplier of the second derivative of the potential, thus the problem is singularly perturbed. We demonstrate the occurence of internal layers at so called device-junctions, which are jump-discontinuities of the data, and present a finite difference scheme which allows for the resolution of these internal layers without employing an exceedingly large number of gridpoints. We establish the relation of this scheme to exponentially fitted schemes and give a convergence proof. Moreover the construction of efficient grids is discussed.

1. Introduction

Since their derivation in the year 1950 by van Roosbroeck [10] the basic semiconductor device equations have been thoroughly investigated by physicists, mathematicians and numerical analysts. In the one-dimensional case they read after appropriate scaling:

$$\left.\begin{array}{lll} (1.1) & \lambda^2\psi'' = n-p-C(x) & \text{Poisson's equation} \\ (1.2) & (\mu_n(n'-n\psi'))' = R & \text{electron continuity equation} \\ (1.3) & (\mu_p(p'+p\psi'))' = R & \text{hole continuity equation} \end{array}\right\} -1 \leq x \leq 1$$

The semiconductor device is assumed to be located in the scaled x-interval [-1,1], ψ denotes the scaled electrostatic potential, n the scaled electron concentration and p the scaled hole concentration. C(x) stands for the scaled doping profile, $\mu_{n,p}$ for the scaled electron (hole) mobility and R for the scaled recombination-generation rate. The scaled electron and hole current densities are given by

(1.4) $J_n = \mu_n(n'-n\psi')$, $J_p = -\mu_p(p'+p\psi')$

and the scaled total current density is

(1.5) $J = J_n + J_p$.

The multiplier λ^2 of the second derivative of the potential is - for modern devices - a small constant. Typically $\lambda^2 \sim 10^{-8}$ holds and therefore (1.1)-(1.3) is singularly perturbed.

Details on the scaling, which leads to the formulation (1.1)-(1.3), on the physical background and on the modelling of the mobilities and of the recombination-generation rate are given in [2], [3], [5], [8].

The maybe most important parameter function of the problem is the doping profile C(x) which determines the performance of the device under consideration (see [8]). Physically, C(x) is a smooth function with a few thin regions of fast variation (so called junctions). For simplicity we will assume for the following that our device has only one junction at, say, x=0, that the doping profile has a jump-discontinuity at the junction, i.e.

(1.6) $C \in C^\infty([-1,0]) \cap C^\infty([0,1]),\ C(0+) \neq C(0-)$

and that

(1.7) $C(x)<0$ on $[-1,0)$ $C(x)>0$ on $(0,1]$

holds. The class of devices we are considering are therefore abrupt pn-diodes,however the generalisation of the following theory to one-dimensional multi-junction devices like fieldeffect transistors and thyristors (see [9]) is straightforward.

We assume that our diode has Ohmic contacts at $x=\pm 1$, which means the following boundary conditions are imposed:

(1.8) $(n-p-C)/_{x=\pm 1} = 0,\ np/_{x=\pm 1} = \delta^4,\ \psi/_{x=\pm 1} = \ln\frac{n}{\delta^2}/_{x=\pm 1} + U_\pm$.

δ^2 denotes the scaled intrinsic number of the semiconductor, which for modern devices is another small parameter. Usually $\delta^2 \sim 10^{-7}$. U_+ and U_- denote the externally applied scaled cathode and anode potential resp. and $U = U_- - U_+$ is the scaled applied voltage.

By a simple calculation we can rewrite (1.8) as Dirichlet boundary conditions for n, p and Ψ:

$$(1.9)\quad n/_{x=\pm 1} = \frac{1}{2}(C+\sqrt{C^2+4\delta^4})/_{x=\pm 1} =: n_{\pm},\ p/_{x=\pm 1} = \frac{1}{2}(-C+\sqrt{C^2+4\delta^4})/_{x=\pm 1} =: p_{\pm}$$

$$(1.10)\quad \psi/_{x=\pm 1} = \ln\left[\frac{C+\sqrt{C^2+4\delta^4}}{2\delta^2}\right]/_{x=\pm 1} + U_{\pm} =: \psi_{\pm}\ .$$

The static device problem (1.1)-(1.3), (1.9), (1.10) constitutes a singularly perturbed two-point boundary value problem with the perturbation parameter λ.

Using matched asymptotic expansions we will demonstrate the occurence of an internal transition layer at the abrupt junction x=0. The solutions n, p and ψ vary exponentially fast within this thin layer and they vary moderately away from x=0. Boundary layers do not occur.

Clearly, the structure of solution has to be taken into account when the applicability of discretisation methods is assessed and analysed. We will show that standard finite difference methods can only be used when a very fine grid is employed within the layer. The main emphasis of this paper lies in the derivation and analysis of an exponentially fitted discretisation method for the continuity equations (1.2), (1.3), which allows for coarse grid and still produces at least "structurally" correct discrete solutions.

We remark that the boundary values $\psi_{\pm}$ blow up as $\delta\to 0$ for fixed $U_{\pm}$. In the following we will however only be interested in λ-asymptotics, i.e. we will assume that δ is constant as λ tends to zero. It can be shown that the so obtained λ-expansions 'depend on δ very weakly' as long as δ not too small compared to λ (note that $\psi_{\pm}$ only blow up logarithmically as $\delta\to 0$) (see [3]).

In Section 2 we will discuss the matched asymptotic expansions; in Section 3 we present an analysis of the convergence of approximate solutions which is based on decoupling the equations. Finite difference schemes and their convergence are discussed in Section 4.

We remark that the one-dimensional semiconductor device problem is practically relevant only for a very restricted class of devices . Its analysis however serves very well as a guideline for understanding the practically important corresponding multi-dimensional problems. It also provides an example for the useful application of exponentially fitted difference schemes, which have not been taken seriously enough by most O.D.E. numerical analysts so far.

2. Matched Asymptotic Expansion

At first we rewrite the problem (1.1)-(1.3) by using new dependent variables. Instead of (ψ,n,p) we introduce (ψ,u,v):

(2.1) $$n=\delta^2 e^{\psi}u, \quad p=\delta^2 e^{-\psi}v.$$

Since n and p are physical carrier concentrations we only admit solutions with n and p positive, i.e. we require

(2.2) $$u>0, \ v>0 \text{ on } [-1,1].$$

The current densities (1.4) are easily computed by

(2.3) $$J_n=\delta^2\mu_n e^{\psi}u', \quad J_p=-\delta^2\mu_p e^{-\psi}v'.$$

For the following we assume that the mobilities μ_n and μ_p are equal to 1 and that the recombination-generation rate R vanishes identically. Then (1.1)-(1.3), (1.9), (1.10) transform to

(2.4) $$\lambda^2\psi'' = \delta^2 e^{\psi}u-\delta^2 e^{-\psi}v-C(x)$$

(2.5) $$(e^{\psi}u')'=0$$

(2.6) $$(e^{-\psi}v')'=0$$

(2.7) $$\psi(\pm 1)=\psi_{\pm}, \quad u(\pm 1)=e^{-U_{\pm}}=:u_{\pm}, \quad v(\pm 1)=e^{U_{\pm}}=:v_{\pm}.$$

The advantage of the set of variables (ψ,u,v) is that the potential ψ appears explicitly in the right hand side of Poisson's equation (2.5). Contrary to (1.1)-(1.3) the system (2.4)-(2.6) is not singular singularly perturbed (see [7]). Also the equations (2.5), (2.6) are self-adjoint and as we will see later on slow and fast solution components are separated. Proceeding as suggested by standard singular perturbation theory we insert the outer expansion

(2.8) $$\begin{pmatrix}\psi(x,\lambda)\\ u(x,\lambda)\\ v(x,\lambda)\end{pmatrix} \sim \sum_{i=0}^{\infty} \lambda^i \begin{pmatrix}\bar{\psi}_i(x)\\ \bar{u}_i(x)\\ \bar{v}_i(x)\end{pmatrix}$$

into (2.4)-(2.6) and obtain equations for the outer terms by equating coefficients of equal λ-powers. Specifically for i=0 we obtain the reduced equations:

(2.9) $0 = \delta^2 e^{\bar\psi_\circ}\bar u_\circ - \delta^2 e^{-\bar\psi_\circ}\bar v_\circ - C(x) \iff \bar\psi_\circ = \ln\left[\frac{C+\sqrt{C^2+4\delta^4\bar u_\circ \bar v_\circ}}{2\delta^2\bar u_\circ}\right]$

(2.10) $(e^{\bar\psi_\circ}\bar u_\circ')' = 0$

(2.11) $(e^{-\bar\psi_\circ}\bar v_\circ')' = 0.$

Since $C(x)$ is discontinuous at $x=0$, at least one of the quantities $\bar\psi_\circ$, $\bar u_\circ$, $\bar v_\circ$ has to be discontinuous at $x=0$ and therefore the full solutions ψ,u,v, which are in $C^1([-1,1])$, cannot be approximated uniformly by $\psi_\circ$, $u_\circ$, $v_\circ$ on $[-1,1]$. Singular perturbation theory predicts the occurence of an internal layer at the junction $x=0$. Near $x=0$ we expect the solutions to vary on the fast scale $\tau=\frac{x}{\lambda}$ and so we are led to setting up the modified expansion

(2.12) $$\begin{pmatrix}\psi(x,\lambda)\\ u(x,\lambda)\\ v(x,\lambda)\end{pmatrix} \sim \sum_{i=0}^{\infty} \lambda^i \cdot \left[\begin{pmatrix}\bar\psi_i(x)\\ \bar u_i(x)\\ \bar v_i(x)\end{pmatrix} + \begin{pmatrix}\hat\psi_i(\tau)\\ \hat u_i(\tau)\\ \hat v_i(\tau)\end{pmatrix}\right].$$

We require the inner terms to satisfy the matching conditions

(2.13) $\hat\psi_i(\pm\infty) = \hat u_i(\pm\infty) = \hat v_i(\pm\infty) = 0.$

By inserting (2.12) into (2.4)-(2.6) and by evaluating close to $x=0$ we obtain equations for the inner terms. A simple calculation shows that $\hat u_\circ \equiv \hat v_\circ \equiv 0$ and therefore

(2.14) $\bar u_\circ(0+) = \bar u_\circ(0-), \quad \bar v_\circ(0+) = \bar v_\circ(0-)$

holds. u and v are slow variables. For $\hat\psi_\circ$ we obtain the layer problem

(2.15) $\ddot{\hat\psi}_\circ = \delta^2 e^{\bar\psi_\circ(0+)+\hat\psi_\circ}\bar u_\circ(0) - \delta^2 e^{-\bar\psi_\circ(0+)-\hat\psi_\circ}\bar v_\circ(0) - C(0+), \quad \tau>0$

(2.16) $\ddot{\hat\psi}_\circ = \delta^2 e^{\bar\psi_\circ(0-)+\hat\psi_\circ}\bar u_\circ(0) - \delta^2 e^{-\bar\psi_\circ(0-)-\hat\psi_\circ}\bar v_\circ(0) - C(0-), \quad \tau<0$

(2.17) $\dot{\hat\psi}_\circ(0+) = \dot{\hat\psi}_\circ(0-), \quad \hat\psi_\circ(0+) - \hat\psi_\circ(0-) = \bar\psi_\circ(0-) - \bar\psi_\circ(0+)$

(2.18) $\hat\psi_\circ(\pm\infty) = 0$

(the dots denote differentiation with respect to τ).
By analysing the first order term of the expansion we find that the current densities J_n and J_p are slow variables, too. Therefore we get

(2.19) $$e^{\bar\psi_\circ(0+)}\bar u_\circ'(0+) = e^{\bar\psi_\circ(0-)}\bar u_\circ'(0-),$$
$$e^{-\bar\psi_\circ(0+)}\bar v_\circ'(0+) = e^{-\bar\psi_\circ(0-)}\bar v_\circ'(0-).$$

Zeroth order layers at the boundaries $x=\pm 1$ do not occur since the boundary data ψ_+, u_+, v_+, satisfy the zero-space-charge-equation (2.9). Thus we impose (2.7) also for the reduced solutions $\bar{\psi}_o$, $\bar{u}_o$, $\bar{v}_o$.

It has been shown in [3] that the reduced problem (2.9) -(2.11), (2.7), (2.14), (2.19) has a solution ($\bar{\psi}_o$, $\bar{u}_o$, $\bar{v}_o$) for all applied potentials $U_+ \in \mathbb{R}$. Also, for given positive $\bar{u}_o(0)$, $\bar{v}_o(0)$ the layer problem (2.15)-(2.18) has a unique exponentially decaying solution Φ_o.

$$(2.20) \qquad |\Phi_o(\tau)| \leq C_1 \exp(-C_2|\tau|); \qquad C_1, C_2 > 0$$

(see [3]).

Collecting the results, we have the asymptotics

$$(2.21)(a) \qquad \psi(x,\lambda) = \bar{\psi}(x) + \Phi(\tfrac{x}{\lambda}) + 0(\lambda)$$

$$(2.21)(b) \qquad u(x,\lambda) = \bar{u}(x) + 0(\lambda)$$

$$(2.21)(c) \qquad v(x,\lambda) = v(x) + 0(\lambda)$$

and by using (2.1)

$$(2.21)(d) \qquad n(x,\lambda) = \bar{n}(x) + \hat{n}(\tfrac{x}{\lambda}) + 0(\lambda)$$

$$(2.21)(e) \qquad p(x,\lambda) = \bar{p}(x) + \hat{p}(\tfrac{x}{\lambda}) + 0(\lambda)$$

uniformly for $x \in [-1,1]$ where $\Phi(\tau)$, $\hat{n}(\tau)$, $\hat{p}(\tau)$ decay to zero exponentially as $\tau \to \pm\infty$. Outside the junction-layer of width $0(\lambda|\ln\lambda|)$ at $x=0$ the solutions are moderately varying functions approximated by the reduced solutions and inside the layer they vary exponentially on the scale $\frac{x}{\lambda}$.

Analogous results for the multidimensional semiconductor device problem can be found in [2], [3].

3. Convergence of Approximations

We now reformulate (2.4)-(2.6) as fixed point problem for ψ. Clearly the continuity equations (2.5), (2.6) subject to the corresponding boundary conditions (2.7) can be solved uniquely for $u=u(\psi)$, $v=v(\psi)$ when $\psi \in C([-1,1])$ is given. We define an operator $R: C([-1,1]) \to C([-1,1])$ by $R(\psi)=\sigma$ where σ is the (unique) solution of

$$(3.1) \qquad \lambda^2\sigma'' = \delta^2 e^{\sigma} u(\psi) - \delta^2 e^{-\sigma} v(\psi) - C(x), \quad -1 \leq x \leq 1$$

(3.2) $\quad \sigma(\pm 1)=\psi_{\pm}$.

Then every solution $(\psi,\ u(\psi),\ v(\psi))$ of the semiconductor device problem (2.4)-(2.7) satisfies

(3.3) $\quad F(\psi):=\psi-R(\psi)=0,\ F:C([-1,1]) \rightarrow C([-1,1])$

and vice versa. We call a solution $(\psi^*,\ u(\psi^*),\ v(\psi^*))$ isolated if the Frechet derivative $D_{\psi}F(\psi^*):C([-1,1]) \rightarrow C([-1,1])$ is boundedly invertible.

The following Theorem is the main tool for the convergence analysis of discretisation methods:

<u>Theorem 3.1</u> Let $(\tilde{\psi},\tilde{u},\tilde{v})\varepsilon(C([-1,1])^3$ satisfy $\tilde{u}(x),\ \tilde{v}(x)\geq e^{\underline{U}}$ for some $\underline{U}\varepsilon\mathbb{R}$ and assume that $D_{\psi}F(\tilde{\psi}):C([-1,1]) \rightarrow C([-1,1])$ is boundedly invertible. Moreover let $\tilde{\psi}_1$ be the (unique) solution of

(3.4) $\quad \lambda^2\tilde{\psi}_1'' = \delta^2 e^{\tilde{\psi}_1}\tilde{u}-\delta^2 e^{-\tilde{\psi}_1}\tilde{v}-C(x),\quad -1\leq x\leq 1$

(3.5) $\quad \tilde{\psi}_1(\pm 1) = \psi_{\pm}$.

Then there is an isolated solution $(\psi^*,\ u^*,\ v^*)$ of (2.4)-(2.7) and there are constants $M_1>0,\ M_2>0,\ \chi>0$ such that

(3.6)
$$\|\psi^*-\tilde{\psi}\|_{[-1,1]} \leq 2M_1(\|\tilde{\psi}-\tilde{\psi}_1\|_{[-1,1]} +M_2\delta^2(\|e^{\tilde{\psi}_1}\|_{[-1,1]}\ \|u(\tilde{\psi})-\tilde{u}\|_{[-1,1]} +\|e^{-\tilde{\psi}_1}\|_{[-1,1]}\ \|v(\tilde{\psi})-\tilde{v}\|_{[-1,1]}))$$

if the right hand side of (3.6) is less than χ.

By going through the proof, which is entirely based on the implicit function theorem and can be found in [6], the constants can easily be identified:

(3.7) $\quad M_1:=\|(D_{\psi}F(\tilde{\psi}))^{-1}\|$

(3.8) $\quad \chi:=\dfrac{1}{(M_3+1)(2M_1+1)}$

where M_3 is the local Lipschitz constant of $D_{\psi}R$ in a neighbourhood of ψ^*. M_2 is bound for the norm of the operator $S:C([-1,1]) \rightarrow C([-1,1])$, where $Sf=w$ is defined by

$$\lambda^2 w''= \delta^2(e^{\xi}u(\tilde{\psi})+e^{-\eta}v(\tilde{\psi}))w+f,\quad -1\leq x\leq 1$$

$$w(\pm 1)=0$$

with $\xi(x)$ and $\eta(x)$ between $\tilde{\psi}_1(x)$ and $R(\tilde{\psi})(x)$.

The theorem allows a decoupled convergence analysis. To show that $(\tilde{\psi},\tilde{u},\tilde{v})$ are close to an isolated solution (ψ^*,u^*,v^*) it suffices to prove that

(i) $\tilde{u}$ and $\tilde{v}$ approximate the solution $u(\tilde{\psi})$, $v(\tilde{\psi})$ of

$$(e^{-\tilde{\psi}}u(\tilde{\psi})')'=0, \quad u(\tilde{\psi})(\pm 1)=u_{\pm}$$

and

$$(e^{-\tilde{\psi}}v(\tilde{\psi})')'=0, \quad v(\tilde{\psi})(\pm 1)=v_{\pm}$$

resp.

(ii) $\tilde{\psi}$ approximates the solution $\tilde{\psi}_1$ of (3.4), (3.5).

(iii) the linearisation of F at $\tilde{\psi}$ is boundedly invertible.

Then the estimate (3.6) holds and

$$\text{(3.9)} \qquad \|\tilde{u}-u^*\|_{[-1,1]} \leq \|\tilde{u}-u(\tilde{\psi})\|_{[-1,1]}+M_4\|\tilde{\psi}-\psi^*\|_{[-1,1]}$$

$$\text{(3.10)} \qquad \|\tilde{v}-v^*\|_{[-1,1]} \leq \|\tilde{v}-v(\tilde{\psi})\|_{[-1,1]}+M_5\|\tilde{\psi}-\psi^*\|_{[-1,1]}$$

follow immediately. M_4 and M_5 are local Lipschitz-constants of the maps $\psi \to u(\psi)$, $\psi \to v(\psi)$ resp. Estimates of the approximation errors of n and p are easily obtained from (2.1).

If $\tilde{\psi}$ is close to ψ^* then

$$\text{(3.11)} \qquad M_1 \sim \|(D_\psi F(\psi^*))^{-1}\| =: M_1^*$$

holds because of the Lipschitz-continuity of $D_\psi F$. The constants M_1^*,M_3,M_4,M_5 are the condition-numbers of the problem (2.4)-(2.7) and are - if the problem is locally well posed - of moderate magnitude even for small λ and δ (this can in fact be shown for a symmetric diode in low injection, i.e. for "not too large" applied voltages U, see [4]). Most importantly, these constants do not depend on the approximations $\tilde{\psi},\tilde{u},\tilde{v}$ (see [4]). Since a closer characterisation of $M_1^*,M_2,\ldots,M_5$ is beyond the scope of this paper we will for the following assume that the problem is - locally at some isolated solution (ψ^*,u^*,v^*) - well posed.

The invertibility of F at the approximation $\tilde{\psi}$ is usually difficult to verify. As mentioned before, if $\tilde{\psi}$ is close to the isolated solution ψ^* then $D_\psi F(\tilde{\psi})$ is boundedly invertible and (3.11) holds. Therefore the practically most relevant assertion of Theorem 3.1 is the estimate (3.6) which can be used to derive error estimates for discretisations which converge to an isolated solution.

4. Finite Difference Methods

To discretise the semiconductor device problem we introduce a grid on [-1,1]:

$$x_0=-1<x_1<\dots<x_{N-1}<x_N=1 \tag{4.1}$$

and define the mesh-sizes

$$h_i=x_{i+1}-x_i. \tag{4.2}$$

For the discrete approximation of $\psi''(x_i)$ we use the three-point-formula. Then the discrete Poisson's equation reads

$$\frac{2\lambda^2}{h_i+h_{i-1}}\cdot\left(\frac{\psi_{i+1}-\psi_i}{h_i}-\frac{\psi_i-\psi_{i-1}}{h_{i-1}}\right)=n_i-p_i-C(x_i),\quad i=1,\dots,N-1 \tag{4.3}$$

The boundary conditions for ψ_i are

$$\psi_0=\psi_-,\quad \psi_N=\psi_+. \tag{4.4}$$

In the sequel we will also employ discrete u and v variables defined analogously to (2.1):

$$n_i=\delta^2 e^{\psi_i}u_i,\quad p_i=\delta^2 e^{-\psi_i}v_i. \tag{4.5}$$

By inserting (4.5) into the right hand side of (4.3) we obtain a discrete version of (2.4).

At this point we remark that the u and v variables are only used for reasons of analytical transparency. For actual computations the set (ψ,n,p) should be employed since the evaluation of exponentials can lead to over- or underflow (particularly for large applied voltages).

The discretisation of the continuity equations has to be done with particular care. To demonstrate why, we assume that the device is in thermal equilibrium, i.e. $U_+=U_-=0$ holds. Then $u\equiv 1$, $v\equiv 1$ solve (2.5), (2.7) and (2.6), (2.7) resp., i.e. $J_n\equiv J_p\equiv 0$ holds. The corresponding equilibrium potential ψ_e is determined by (2.4), (2.7) with u and v equal to 1. A typical equilibrium potential is depicted in Figure 1.

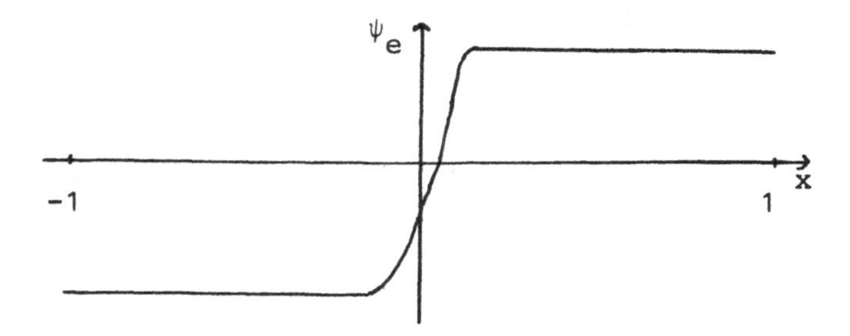

Figure 1: Equilibrium Potential

The electron continuity equation (1.2) reduces to the initial value problem ($\mu_n \equiv 1$, $R \equiv 0$ is assumed!)

(4.6) $\quad n' = n\psi_e', \quad -1 \leq x \leq 1, \quad n(-1) = n_-$

with the equilibrium electron concentration $n_e = \delta^2 e^{\psi_e}$ as solution (see Figure 2).

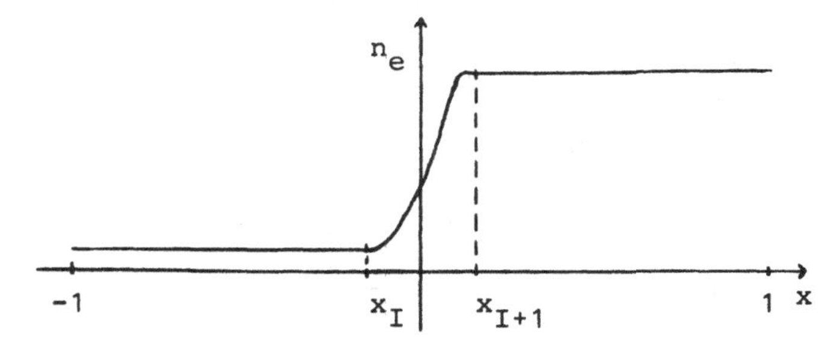

Figure 2: Equilibrium Electron Concentration

For given ψ_e we discretise (4.6) by the following simple method

$$(4.7) \qquad \frac{n_{i+1}-n_i}{h_i} = \frac{n_{i+1}+n_i}{2} \; \frac{\psi_e(x_{i+1})-\psi_e(x_i)}{h_i}$$

whose solution is given by

$$(4.8) \qquad n_i = \prod_{j=0}^{i-1} \Phi\left(\frac{\psi_e(x_{j+1})-\psi_e(x_j)}{2}\right) n_-, \qquad \Phi(z) = \frac{1+z}{1-z}$$

Assuming that x_I is the largest grid-point to the left of the layer, that no grid-point is placed inside the layer and that x_{I+1} is the smallest grid-point to the right of the layer (see Figure 2) we obtain

$$(4.9) \qquad n_I \sim n_e(x_I) = \delta^2 e^{\psi_e(x_I)}$$

if the grid is sufficiently fine in $[-1,x_I]$ and

$$(4.10) \qquad n_{I+1} = n_I \Phi\left(\frac{\psi_e(x_{I+1})-\psi_e(x_I)}{2}\right) \neq n_e(x_{I+1}) = \delta^2 e^{\psi_e(x_{I+1})}$$

since $\Phi(z)$ does not approximate e^{2z} unless $z\sim 0$. Our simple trapezoidal-type scheme does not resolve the layer-jump correctly unless a fine grid is used inside the junction layer.

To derive an uniformly convergent discretisation of (2.5) we denote by ψ_h the piecewise linear interpolant of ψ:

$$(4.11) \qquad \psi_h(x) = \psi_i + (x-x_i)\,\frac{\psi_{i+1}-\psi_i}{h_i}, \quad x_i \leqq x \leqq x_{i+1}$$

and approximate (2.5) by the two-point-boundary-value problem

$$(4.12) \qquad (e^{\psi_h} u_h')'=0, \quad u_h(-1)=u_-, \quad u_h(1)=u_+.$$

Integrating (4.12) and multiplying by $e^{-\psi_h}$ gives

$$u_h' = \text{const.}\, e^{-\psi_h}.$$

Another integration, now from x_i to x_{i+1}, gives

$$(4.13) \qquad u_{i+1}-u_i = \text{const.}\ \frac{e^{-\psi_{i+1}}-e^{-\psi_i}}{\psi_i-\psi_{i+1}}\, h_i$$

where we denoted $u_i := u_h(x_i)$. (4.13) can be rearranged as a finite difference scheme for (2.5), (2.7):

$$(4.14) \qquad \frac{2}{h_i+h_{i-1}}\cdot\left(\frac{\psi_i-\psi_{i+1}}{e^{-\psi_{i+1}}-e^{-\psi_i}}\cdot\frac{u_{i+1}-u_i}{h_i}-\frac{\psi_{i-1}-\psi_i}{e^{-\psi_i}-e^{-\psi_{i-1}}}\cdot\frac{u_i-u_{i-1}}{h_{i-1}}\right)=0,$$

$$i=1,\ldots,N-1$$

$$(4.15) \qquad u_0=u_-, \quad u_N=u_+.$$

Resubstituting $n_i=\delta^2 e^{\psi_i} u_i$ gives (after a lengthy calculation)

$$(4.16) \qquad \frac{2}{h_i+h_{i-1}}\Big[\Big(\gamma\Big(\frac{\psi_{i+1}-\psi_i}{2}\Big)\cdot\frac{n_{i+1}-n_i}{h_i}-\frac{\psi_{i+1}-\psi_i}{h_i}\cdot\frac{n_{i+1}+n_i}{2}\Big)$$

$$-\Big(\gamma\Big(\frac{\psi_i-\psi_{i-1}}{2}\Big)\cdot\frac{n_i-n_{i-1}}{h_{i-1}}-\frac{\psi_i-\psi_{i-1}}{h_{i-1}}\cdot\frac{n_i+n_{i+1}}{2}\Big)\Big]=0,$$

$$i=1,\ldots,N-1$$

(4.17) $n_0=n_-$, $n_N=n_+$

where we set

(4.18) $\gamma(z)=z\cdot\coth z$.

An analogous discretisation can be derived for the hole-continuity equation (1.3).

(4.14), (4.15) resp. (4.16), (4.17) is the celebrated Scharfetter-Gummel (SG) scheme, which has been used for about 20 years for the numerical solution of the basic semiconductor device equations (particularly its multi-dimensional analogue, see [3]) without complete mathematical understanding.

The SG-scheme is distinguished from the trapezoidal-type scheme by the function $\gamma(z)$ which appears - with appropriate arguments - as a factor of the approximations of n'. Clearly $\gamma(0)=1$ holds, thus (4.16) is approximately equal to the trapezoidal scheme in those regions where ψ_{i+1} is close to ψ_i, i.e. away from the junction layer.
The SG-scheme

$$\gamma\left(\frac{\psi(x_{i+1})-\psi(x_i)}{2}\right)\cdot\frac{n_{i+1}+n_i}{2} = \frac{n_{i+1}+n_i}{2}\cdot\frac{\psi(x_{i+1})-\psi(x_i)}{h_i}$$

integrates the initial value problem

$$n'=n\psi'$$

exactly, therefore the factor $\gamma(z)$ allows for the accurate resolution of the layer jump of n indepently of the meshsizes inside the layer.

Also, another interpretation of the SG-scheme is possible. (4.14) can be regarded as exponentially fitted difference scheme for (2.5) (see [1]) with the fitting factor $\gamma(z)$. The special form (4.18) of the fitting factor can be derived by the methods given in [1], that is by choosing $\gamma(z)$ such that the resulting scheme is uniformly (in λ) convergent assuming that the asymptotic expansions (2.21a,d) hold for ψ and n.

The existence of discrete solutions $(\psi_i,u_i,v_i)_{i=0}^N$ of (4.3), (4.4), (4.14), (4.15) and of the SG-discretisation of the hole-continuity equation is shown by rewriting the scheme as the fixed point problem $\psi_h=R_h(\psi_h)$ where ψ_h denotes the piecewise linear interpolant of $(\psi_i)_{i=0}^N$ and R_h is the discrete analogue of the operator R defined at the beginning of Section 3. The discrete maximum principle gives upper and lower bounds for ψ_h and Schauder's Theorem implies the existence of a fixed point of the operator R_h (note that $R_h:\mathbb{R}^{N+1}\to\mathbb{R}^{N+1}$).

To apply Theorem 2.1 to the convergence analysis we set $\tilde{\psi}:=\psi_h$, $\tilde{u}=u_h$, $\tilde{v}=v_h$ where ψ_h is a fixed point of R_h, u_h solves (4.12) and v_h is the solution of

$$(4.19)\qquad (e^{-\psi_h}v_h')'=0,\quad v_h(\pm 1)=v_\pm$$

$u_i:=u_h(x_i)$, $v_i:=v_h(x_i)$ are the discrete approximations of $u(x_i)$, $v(x_i)$.

Since $\tilde{u}=u(\psi_h)$, $\tilde{v}=v(\psi_h)$ holds we obtain the error estimate

$$(4.20)\qquad \| \psi_h-\psi^*\|_{[-1,1]} \leq 2M_1 \| \psi_h-\psi_{h,1}\|_{[-1,1]}$$

at an isolated solution ψ^*, where $\psi_{h,1}$ solves

$$(4.21)\qquad \lambda^2\psi_{h,1}'' = \delta^2 e^{\psi_{h,1}}u_h-\delta^2 e^{-\psi_{h,1}}v_h-C(x),\quad -1\leq x\leq 1$$

$$(4.22)\qquad \psi_{h,1}(\pm 1)=\psi_\pm .$$

Since (4.3), (4.4), (4.5) is a uniformly stable discretisation for (4.21), (4.22) (see [2]) the estimate

$$(4.23)\qquad \| \psi_h-\psi_{h,1}\|_{[-1,1]} \leq C_s \max\{|l_i|, i=1,\dots,N-1\}$$

follows, where $(l_i)_{i=1}^{N-1}$ is the local discretisation error obtained by inserting $\psi_{h,1}$ into (4.3) (using (4.5)) and C_s is a constant independent of λ. C_s may depend weakly on δ, viz. usually $C_s\sim D|\ln\delta|$, $D>0$.

Taylor expansion gives

$$(4.24)\qquad |l_i|\leq C_L\lambda^2 h_i \max\{|\psi_{h,1}'''(x)|, x\in[x_{i-1},x_{i+1}]\}$$

if $C_1\leq\frac{h_i}{h_{i-1}}\leq C_2$ holds. The estimate can be improved for quasi-uniform grids. Then $|l_i| = O(h_i^2)$ holds.

An asymptotic analysis analogous to that in Section 2 shows that $\psi_{h,1}$ has an exponential layer at x=0, i.e.

$\psi_{h,1}=\bar{\psi}_{h,1}(x)+\Phi_{h,1}(\frac{x}{\lambda})+0(\lambda)$ holds and therefore we obtain within the layer

$$(4.25) \qquad |l_i|\le C_L\frac{h_i}{\lambda}\max|\dddot{\Phi}_{h,1}(\frac{x}{\lambda})|\le D\frac{h_i}{\lambda}\exp\left(-\frac{C_3|x_i|}{\lambda}\right);D,C_3>0.$$

When equidistributing l_i, i.e. when we choose h_i such that the right side of (4.25) is equal to some error bound ω, we obtain a grid which is coarse outside the layer and fine inside the layer.

$$(4.26) \quad \begin{cases} h_i=0(\lambda\omega\exp(\frac{C_3|x_i|}{\lambda})) \text{ for } |x_i|\le\frac{1}{C_3}\lambda|\ln\lambda| \\ h_i=0(\omega) \text{ for } |x_i|\ge\frac{1}{C_3}\lambda|\ln\lambda| \end{cases}$$

and the error estimate

$$(4.27) \qquad \max_i|\psi_i-\psi^*(x_i)|=0(\omega)$$

holds because of (4.20), (4.23). Grids satisfying (4.26) can be chosen such that they have $0(\frac{1}{\omega})$ grid points.

Estimates for $u_i-u^*(x_i)$ and $v_i-v^*(x_i)$ immediately follow from (3.9), (3.10).

Since the equidistributing grid (4.26) is fine within the layer, what is now the advantage of the SG-method over, say, the trapezoidal type method? Firstly, the approximation error of the discretisation scheme involving the SG-discretisation is small, if only the local discretisation error of Poisson's equation is equidistributed (with a suff. small error bound ω). The local error of the continuity equations can be disregarded. For some applications (particularly higher dimensional problems) however it is undesirable to fully equidistribute the local error of Poisson's equation since it might require an exceedingly large amount of mesh-points. The second advantage of the SG-scheme is that the discretisation of the semiconductor problem (4.3), (4.4), (4.14), (4.15), (4.19) gives 'structurally correct' solutions independently of the grid, i.e. the layer structure of the solutions is retained, the layer jump is resolved correctly and large errors only occur within the layer no matter how the grid is chosen.

To demonstrate this we choose a layer ignoring grid, i.e. $h_i \gg \lambda$ for all i, and again do a decoupled analysis. Then we obtain from (4.3), (4.5)

$$(4.28) \quad \psi_i = \ln\left[\frac{C(x_i)+\sqrt{C(x_i)^2+4\delta^4 u(x_i)v(x_i)}}{2\delta^2}\right]+O\left(\max_i \frac{\lambda}{h_i^2}\right).$$

Thus $\psi_i \sim \bar{\psi}_o(x_i)$. The reduced solution $\bar{\psi}_o$ is resolved accurately on a layer ignoring grid (see (2.9)). A more difficult calculation shows that the discrete solutions of Poisson's equation are accurate outside the layer even if $h_i \sim \lambda$ is chosen (see [2]).

To analyse the error of the SG-scheme we apply a standard estimate from the theory of elliptic equations to (4.12) and (2.5), (2.7) and obtain

$$(4.29) \quad \| u_h - u^* \|_{H^1(-1,1)} \leq \text{const.} \| \psi_h - \psi^* \|_{L^2(-1,1)}$$

where - in the decoupled analysis - ψ_h denotes the linear interpolant of the exact solution ψ^*. Since the L^2-norm of the error appears on the right hand side we derive by using Sobolov's imbedding Theorem

$$(4.30) \quad \max_i |u_i - u^*(x_i)| \leq \text{const.} (\max_i h_i^2 + \lambda |\ln\lambda|)$$

independently of the grid.

Finally we note that also the current densities J_n and J_p are resolved accurately by the SG-method. A simple calculation gives

$$(4.30) \quad |J^* - J_h| \leq \text{const.} \| \psi^* - \psi_h \|_{L^2(-1,1)}$$

where $J^* \equiv \delta^2 e^{\psi^*} u^{*\prime} - \delta^2 e^{-\psi^*} v^{*\prime}$ is the exact total current density and $J_h \equiv \delta^2 e^{\psi_h} u_h' - \delta^2 e^{-\psi_h} v_h'$ its SG-approximation. J_h can be easily expressed in terms of point values ψ_i, u_i and v_i. An obvious drawback of the SG-scheme is that it is only convergent of first order on non-quasi-uniform grids This, however is compensated by the uniform convergence, which is retained by its multidimensional analogue.

References

[1] E.P.Doolan, J.J.H.Miller and W.H.A.Schilders, "Uniform Numerical Methods for Problems with Initial and Boundary Layers", Boole Press, 1980

[2] P.A.Markowich, C.A.Ringhofer, S.Selberherr and M.Lentini, "A Singular Perturbation Approach for the Analysis of the Fundamental Semiconductor Equations", IEEE Trans.Electron.Devices, Vol.ED-30. No.9, pp.1165-1180, 1983

[3] P.A.Markowich, "A Singular Perturbation Analysis of the Fundamental Semiconductor Device Equations", to appear in SIAM J.Appl.Math., 1984

[4] P.A.Markowich and C.Schmeiser, "Asymptotic Representation of Solutions of the Basic Semiconductor Device Equations", to appear as MRC-TSR, 1984

[5] P.A.Markowich and C.A.Ringhofer, "A Singularly Perturbed Boundary Value Problem Modelling a Semiconductor Device", SIAM J.Appl.Math., Vol.44, No.3, pp.231-256, 1984

[6] M.Mock, " Analysis of Mathematical Models of Semiconductor Devices", Boole Press, 1983

[7] C.Schmeiser and R.Weiss, "Asymptotic Analysis of Singularly Perturbed Boundary Value Problems", submitted to SIAM J.Math.Anal., 1984

[8] S.Selberherr, "Analysis and Simulation of Semiconductor Devices", Springer Verlag, 1984

[9] S.M.Sze, "Physis of Semiconductor Devices", Wiley Interscience, 1981

[10] W.V.Van Roosbroeck,"Theory of Flow of Electrons and Holes in Germanium and other Semiconductors", Bell. Syst.Tech.J., Vol.29, pp.560-607, 1950

Progress in Scientific Computing, Vol. 5
Numerical Boundary Value ODEs

SOLUTION OF PREMIXED AND COUNTERFLOW DIFFUSION FLAME PROBLEMS BY ADAPTIVE BOUNDARY VALUE METHODS

Mitchell D. Smooke[1]
James A. Miller[2]
Robert J. Kee[3]

1. Introduction

Combustion models that simulate pollutant formation and study chemically controlled extinction limits in flames often combine detailed chemical kinetics with complicated transport phenomena. Two of the simplest models in which these processes are studied are the premixed laminar flame and the counterflow diffusion flame. In both cases the flow is essentially one-dimensional and the governing equations can be reduced to a set of coupled nonlinear two-point boundary value problems with separated boundary conditions.

Laminar premixed flames are obtained in the laboratory by flowing a premixed fuel and oxidizer through a cooled porous plug burner. As the gases emerge from the burner they are ignited and a steady flame sits above the burner surface (see Figure 1.1). In general, there is a positive temperature gradient at the burner surface so the problem is nonadiabatic. In the adiabatic case, the flame sits an infinite distance downstream with a zero temperature gradient at the burner surface. The premixed flame problem is appealing due to its simple flow geometry and it has been used by kineticists in understanding elementary reaction mechanisms in the oxidation of fuels [10-11].

Counterflow diffusion flames have been used by combustion scientists to help determine the chemically controlled extinction limits of diffusion flames [2,12,22-25]. Experimentally a reaction zone is stabilized near the stagnation point of two infinitely wide counterflowing coaxial concentric jets. Several modifications of this basic configuration have been investigated and the one we consider in this paper has been used, for example, by Tsuji [25] and Tsuji and Yamaoka [22-24]. Physically, fuel is blown from a porous cylinder into an oncoming stream of air (see Figure 1.2). A free stagnation line parallel to the cylinder axis forms in front of the cylinder's porous surface. Combustion occurs within a thin flame zone near the stagnation line where the fuel and oxidizer are in stoichiometric proportion.

Both problems are posed most naturally as nonlinear two-point boundary value problems on semi-infinite intervals. Despite their outwardly simple form, it has only been

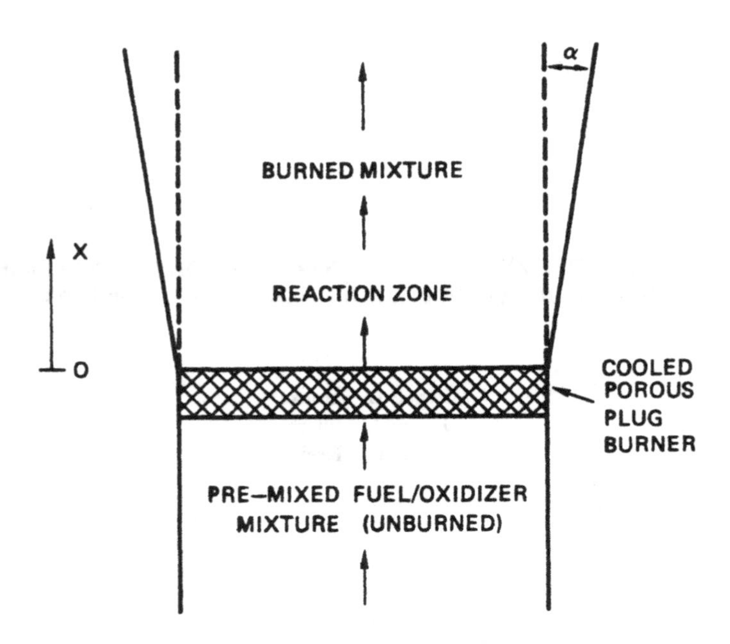

Figure 1.1 Illustration of the premixed laminar flame. In general, the flame gases expand (expansion angle α) as a function of the height x above the burner.

within the last five years that they have been solved with adaptive boundary value solution techniques. Until then, both problems were solved, almost without exception, by time-dependent relaxation methods. This is understandable when one considers the fact that the governing equations of both problems are very sensitive to changes in the boundary data. As a result, the use of initial value techniques like shooting or multiple shooting is not feasible. Instead a global method such as finite differences must be used. However, reliable adaptive global boundary value problem software has only been developed within the last decade.

In this paper we calculate both premixed and counterflow diffusion flame structure by adaptive boundary value methods. In the next section the two models are formulated as nonlinear two-point boundary value problems and in Section 3 the method of solution is discussed. In Section 4 we present numerical results for an ozone-oxygen premixed flame and a methane-air counterflow diffusion flame.

2. Problem Formulation

The formulation of the premixed flame problem we consider closely follows the one originally proposed by Hirschfelder and Curtiss [5]. Our goal is to predict theoretically the mass fractions of the species and the temperature as functions of the independent coordinate x above the burner surface. If the problem is adiabatic, we also want to calculate the mass flow rate since, in general, this quantity is not known in such problems. Upon neglecting viscous effects, body forces, radiative heat transfer and the diffusion of heat due to concentration gradients, the equations governing the structure of a steady one-dimensional isobaric flame (with expansion angle $\alpha = 0$) are

$$\dot{M} = \rho u = \text{constant}, \tag{2.1}$$

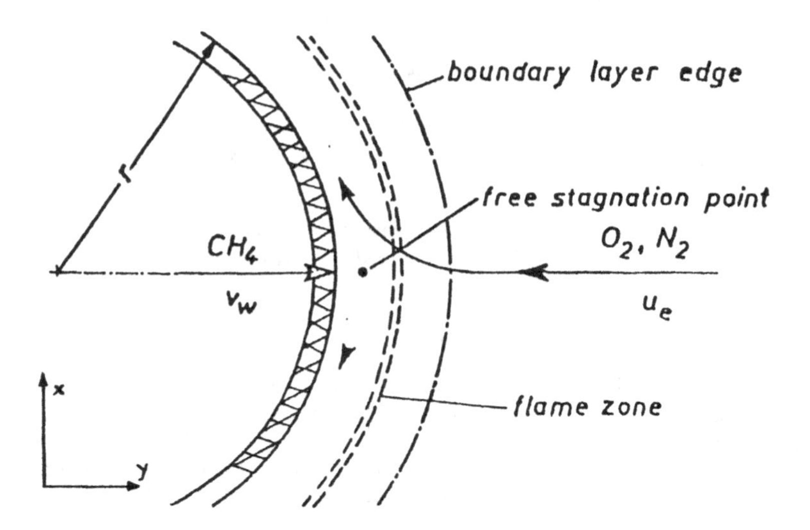

Figure 1.2 Illustration of the counterflow diffusion flame flow configuration.

$$\dot{M}\frac{dY_k}{dx} = -\frac{d}{dx}(\rho Y_k V_k) + \dot{\omega}_k W_k, \qquad k = 1, 2, \ldots, K, \tag{2.2}$$

$$\dot{M}\frac{dT}{dx} = \frac{1}{c_p}\frac{d}{dx}\left(\lambda\frac{dT}{dx}\right) - \frac{1}{c_p}\sum_{k=1}^{K}\rho Y_k V_k c_{p_k}\frac{dT}{dx} - \frac{1}{c_p}\sum_{k=1}^{K}\dot{\omega}_k h_k W_k, \tag{2.3}$$

$$\rho = \frac{p\overline{W}}{RT}. \tag{2.4}$$

In these equations x denotes the independent spatial coordinate; $\dot{M}$, the mass flow rate; T, the temperature; Y_k, the mass fraction of the kth species; p, the pressure; u, the velocity of the fluid mixture; ρ, the mass density; W_k, the molecular weight of the kth species; $\overline{W}$, the mean molecular weight of the mixture; R, the universal gas constant; λ, the thermal conductivity of the mixture; c_p, the constant pressure heat capacity of the mixture; c_{pk}, the constant pressure heat capacity of the kth species; $\dot{\omega}_k$, the molar rate of production of the kth species per unit volume; h_k, the specific enthalpy of the kth species; and V_k, the diffusion velocity of the kth species. The diffusion velocity incorporates Fickian diffusion and thermal diffusion effects. The detailed form of the chemical production rates are given in Kee, Miller, and Jefferson [7] and the details of the transport coefficients can be found in Kee, Warnatz, and Miller [8].

The boundary conditions for the problem are given by

$$T(0) = T_b, \tag{2.5}$$

$$Y_k(0) + \frac{\rho Y_k(0) V_k}{\dot{M}} = \epsilon_k, \quad k = 1, 2, \ldots, K, \tag{2.6}$$

$$\frac{dT}{dx} = 0, \quad \text{as } x \to \infty, \tag{2.7}$$

$$\frac{dY_k}{dx} = 0, \quad \text{as} \quad x \to \infty, \quad k = 1, 2, \ldots, K, \tag{2.8}$$

where T_b is the burner temperature and ϵ_k is the known incoming mass flux fraction of the kth species. We point out that instead of solving the governing equations on the semi-infinite domain $0 \leq x < \infty$, we pose the problem on the finite interval $0 \leq x \leq L$ where the choice of L must be large enough to ensure that the zero flux boundary conditions are "satisfied".

If the problem is adiabatic, the mass flow rate $\dot{M}$ is not known; it is an eigenvalue to be determined. Calculation of the flow rate proceeds by introducing the trivial differential equation

$$\frac{d\dot{M}}{dx} = 0, \tag{2.9}$$

and an additional boundary condition to the system in (2.1-2.8). The particular choice of the extra boundary condition is somewhat arbitrary. It must be chosen, however, to insure that the spatial gradients of both the temperature and the mass fractions are vanishingly small at $x = 0$. We have found that since the problem is most sensitive to variations in the temperature, it is best to fix the temperature at an internal point. We specify

$$T(x_f) = T_f, \tag{2.10}$$

where x_f is a specified spatial coordinate interior to the domain and T_f is a specified temperature. Values of x_f and T_f should be chosen to guarantee a zero temperature gradient at the burner surface. In practice, we can only approach the zero gradient asymptotically and our experience has shown that the optimum position at which to fix the temperature is just past the transition point where the temperature begins to rise rapidly. A good rule of thumb is to choose T_f about 100K above the unburned gas temperature and x_f about one "flame thickness" from the cold boundary.

Our model for the counterflow diffusion flame assumes the flow in the neighborhood of the cylinder (see Figure 1.2) to be laminar, plane stagnation point flow. Hence, the governing boundary layer equations for mass, momentum, chemical species and energy can be written in the form

$$\frac{\partial(\rho u)}{\partial x} + \frac{\partial(\rho v)}{\partial y} = 0, \tag{2.11}$$

$$\rho u \frac{\partial u}{\partial x} + \rho v \frac{\partial u}{\partial y} + \frac{\partial P}{\partial x} = \frac{\partial}{\partial y}\left(\mu \frac{\partial u}{\partial y}\right), \tag{2.12}$$

$$\rho u \frac{\partial Y_k}{\partial x} + \rho v \frac{\partial Y_k}{\partial y} = \dot{\omega}_k W_k - \frac{\partial}{\partial y}\left(\rho Y_k V_{ky}\right), \tag{2.13}$$

$$\rho u c_p \frac{\partial T}{\partial x} + \rho v c_p \frac{\partial T}{\partial y} = \frac{\partial}{\partial y}\left(\lambda \frac{\partial T}{\partial y}\right) - \sum_{k=1}^{K} \rho Y_k V_{ky} c_{pk} \frac{\partial T}{\partial y} - \sum_{k=1}^{K} \dot{\omega}_k W_k h_k. \tag{2.14}$$

The system is closed with the ideal gas law,

$$\rho = \frac{p\overline{W}}{RT}. \tag{2.15}$$

In addition to the variables already defined, μ is the viscosity of the mixture; V_{ky} is the diffusion velocity in the y direction and u and v are the tangential and the radial components of the velocity, respectively. The free stream (tangential) velocity at the edge of the boundary layer is given by $u_e = ax$ where a is the flow strength or velocity gradient.

Upon introducing the similarity variable

$$\eta = \sqrt{\frac{a}{(\rho\mu)_e}} \int_0^y \rho \, dy, \tag{2.16}$$

and the quantities

$$f' = \frac{u}{u_e}, \tag{2.17}$$

$$V = \frac{\rho v}{\sqrt{(\rho\mu)_e a}}, \tag{2.18}$$

the boundary layer equations can be transformed into a system of ordinary differential equations valid along the stagnation-point streamline $x = 0$. We obtain

$$\frac{\partial V}{\partial \eta} + f' = 0, \tag{2.19}$$

$$(f')^2 + V\frac{\partial f'}{\partial \eta} - \frac{\rho_e}{\rho} = \frac{\partial}{\partial \eta}\left(\frac{\rho\mu}{(\rho\mu)_e}\frac{\partial f'}{\partial \eta}\right), \tag{2.20}$$

$$V\frac{\partial Y_k}{\partial \eta} - \frac{\dot{\omega}_k W_k}{\rho a} + \frac{1}{\sqrt{(\rho\mu)_e a}}\frac{\partial}{\partial \eta}(\rho Y_k V_k) = 0, \tag{2.21}$$

$$V c_p \frac{\partial T}{\partial \eta} = \frac{1}{(\rho\mu)_e}\frac{\partial}{\partial \eta}\left(\rho\lambda\frac{\partial T}{\partial \eta}\right) - \frac{1}{\sqrt{(\rho\mu)_e a}}\sum_{k=1}^{K}\rho Y_k V_k c_{pk}\frac{\partial T}{\partial \eta} - \frac{1}{\rho a}\sum_{k=1}^{K}\dot{\omega}_k W_k h_k, \tag{2.22}$$

where the index e refers to quantities at the edge of the boundary layer (free stream values). The boundary conditions at $\eta = 0$ are given by

$$f'(0) = 0, \tag{2.23}$$

$$V(0) = V_w, \tag{2.24}$$

$$Y_k(0) + \frac{\rho Y_k(0) V_k}{V_w\sqrt{(\rho\mu)_e a}} = \epsilon_k, \quad k = 1, 2, \ldots, K, \tag{2.25}$$

$$T(0) = T_w, \tag{2.26}$$

and as $\eta \to \infty$ by

$$f' = 1, \tag{2.27}$$

$$Y_k = Y_{ke}, \quad k = 1, 2, \ldots, K, \tag{2.28}$$

$$T = T_e. \tag{2.29}$$

The wall velocity and temperature (V_w and T_w) are given, as are the mass fractions of the species and the temperature (Y_{ke} and T_e) at the edge of the boundary layer.

3. Method of Solution

The solution of the governing equations in (2.1-2.10) and (2.17-2.29) proceeds by an adaptive finite difference method. In particular, on the mesh $\mathcal{M}$

$$\mathcal{M} = \{0 = x_0 < x_1 < \ldots < x_M = L\}, \tag{3.1}$$

we approximate the spatial operators in the governing equations by finite difference expressions. Diffusion terms are approximated by centered differences and convective terms by upwind approximations. The problem of finding an analytic solution to the equations is then converted into one of finding an approximation to this solution at each point x_j of the mesh $\mathcal{M}$. With the difference equations written in residual form, we seek the solution z^* of the system on nonlinear equations

$$F(z^*) = 0. \tag{3.2}$$

For an initial solution estimate z^0 which is sufficiently close to z^*, the system of nonlinear equations in (3.2) can be solved by Newton's method. This leads to the iteration

$$J(z^n)(z^{n+1} - z^n) = -\lambda^n F(z^n), \quad n = 0, 1, 2, \ldots. \tag{3.3}$$

The iteration continues until the size of $\|z^{n+1} - z^n\|$ is reduced appropriately. $J(z^n) = \partial F(z^n)/\partial z$ is the Jacobian matrix and λ^n $(0 < \lambda \leq 1)$ is the nth damping parameter. For premixed flames the vector z is composed of the dependent variables T and Y_k, and, if the problem is adiabatic, it also includes the mass flow rate $\dot{M}$. In the counterflow diffusion flame, the dependent variable vector is composed of the quantities V, f', Y_k and T. In both problems the block tridiagonal Newton equations are solved with Hindmarsh's subroutines DECBT and SOLBT [4].

The solution of the governing equations in both the premixed and counterflow problems contains regions in which the dependent variables exhibit high spatial activity (steep fronts and sharp peaks). Efficient solution of these problems requires that the high activity regions be resolved adaptively. Techniques that attempt to equidistribute positive weight functions have been used with a great deal of success in these types of problems. Today flames with 30 to 40 chemical species and over 100 chemical reactions can be solved efficiently by adaptively placing grid points in the high activity regions [10-11,14-15,17].

Once a solution has been obtained on the coarsest grid, we determine a new grid by equidistributing a positive weight function over consecutive intervals. For a weight function W and constant C, we determine the grid points such that

$$\int_{x_j}^{x_{j+1}} W \, dx < C, \quad j = 1, 2, \ldots, M-1. \tag{3.4}$$

The weight function is chosen such that the grid points are placed in regions of high spatial activity with the goal of reducing the local discretization error. We use a combination of first and second derivatives of the solution profiles as the weight function. If in any mesh interval there is too large a change in any component of the solution or its slope, then a mesh point is inserted at the midpoint of that interval. In our implementation, the particular combinations of function and slope and the value C can be changed in order to produce a solution of desired accuracy. Typically grids with 40 to 60 adaptively chosen points are required to obtain accurate flame profiles. Our results indicate that as many as 600-800 equispaced points, however, may be needed to achieve equivalent accuracy (see [16]).

In both the premixed and counterflow problems the formation and factorization of the Jacobian matrix accounts for over 90 percent of the cost of the calculation. As a result, the use of a modified Newton method in which the Jacobian is only periodically re-evaluated is indicated. The immediate implication of applying the modified Newton method is that the LU factorization of the Jacobian can be stored and each modified Newton iteration can be obtained by performing relatively inexpensive back substitutions. The problem one faces when applying the modified method is how to determine whether the rate of convergence is fast enough. If the rate is too slow we want to change back to a full Newton method and make use of new Jacobian information. If the rate of convergence is acceptable, we want to continue performing modified Newton iterations.

In our adaptive grid strategy the equidistribution condition is checked one mesh interval at a time and grid points are added appropriately. The coarse grid solution is then interpolated linearly onto the new finer grid. The interpolated result serves as an initial solution estimate for the iteration procedure on the finer grid. The process is continued on successively finer and finer grids until several termination criteria are satisfied. We anticipate that as the size of the mesh spacing gets smaller, the interpolated solution should become a better starting estimate for Newton's method on the next finer grid. For a class of nonlinear two-point boundary value problems Smooke and Mattheij [19] have shown that there exists a critical mesh spacing such that the interpolated solution lies in the domain of convergence of Newton's method on the next grid. As a result, the hypotheses of the Kantorovich theorem [6] are satisfied and the sequence of successive modified Newton iterates can be shown to satisfy a recurrence relation scaled by the first Newton step [18]. We have

Lemma 3.1 If the hypotheses of the Kantorovich theorem are satisfied, the sequence of successive modified Newton estimates can be bounded by

$$\| x^{n+1} - x^n \| \leq c_n \| x^1 - x^0 \|, \quad n = 1, 2, \ldots, \tag{3.5}$$

where c_n is a constant that depends only on the iteration number. •

If in the course of a calculation, we determine that the size of the $(n+1)^{st}$ modified Newton step is larger than the right-hand side in (3.5), we form a new Jacobian and restart the iteration count. We have found that the error estimate in Lemma 3.1 can reduce the cost of the flame calculations by as much as 50 percent over a full Newton method.

The coarse to fine grid strategy eliminates many of the convergence difficulties associated with solving the governing equations directly. However, convergence of Newton's method on the initial grid requires a "good" initial estimate for z^0. We can obtain such a solution estimate by applying a time-dependent starting method on the initial grid. We remark that fundamentally there are two mathematical approaches for solving one-dimensional flame problems - one uses a transient method and the other solves the steady-state boundary value problem directly. Generally speaking, the transient methods are robust but computationally inefficient compared to the boundary value methods, which are efficient but have less desirable convergence properties. Most of the numerical techniques that have been used to solve flame problems have employed a time-dependent method. Variations of this approach have been considered by a variety of researchers (see, e. g., [1,9,20,26]). In these methods, the original nonlinear two-point boundary value problem is converted into a nonlinear parabolic mixed initial-boundary value problem. This is accomplished by appending the term $\partial(\bullet)/\partial t$ to the left-hand side of the conservation equations. This results in a semidiscrete set of equations

$$\frac{\partial z}{\partial t} = F(z), \tag{3.6}$$

with appropriate initial conditions. If the time derivative is replaced, for example, by a backward Euler approximation, the governing equations can be written in the form

$$\mathcal{F}(z^{n+1}) = F(z^{n+1}) - \frac{(z^{n+1} - z^n)}{\tau^{n+1}} = 0, \tag{3.7}$$

where for a function $g(t)$ we define $g^n = g(t^n)$ and where the time step $\tau^{n+1} = t^{n+1} - t^n$. At each time level we must solve a system of nonlinear equations that look very similar to the nonlinear equations in (3.2). Newton's method can again be used to solve this system. The important difference between the system in (3.2) and (3.7) is that the diagonal of the steady-state Jacobian is weighted by the quantity $1/\tau^{n+1}$. This produces a better conditioned system and the solution from the nth time step ordinarily provides an excellent starting guess to the solution at the $(n+1)^{\text{st}}$ time level. The work per time step is similar to that for the modified Newton iteration, but the timelike continuation of the numerical solution produces an iteration strategy that will, in general, be less sensitive to the initial starting estimate than if Newton's method were applied to (3.2) directly. We have found that in a number of flame problems, 10-20 time steps often produce a solution approximation that lies in the domain of convergence of Newton's method on the initial grid. As a result, when we ultimately implement Newton's method on the steady-state equations directly, we obtain a converged numerical solution with only a few additional iterations. This time-dependent starting procedure can also be used on grids other than the initial one.

Once a solution to the flame equations has been obtained, it is often useful to investigate those parameters to which the solution is most "sensitive". Sensitivity analysis is an important tool for the physical investigation and validation of mathematical models (see, e. g., [3]). In the past, however, the major obstacle in obtaining sensitivity information systematically was the additional amount of computation required for solving the sensitivity equations. An advantage of solving the premixed and counterflow flames by Newton's method is that first order sensitivity information can be obtained at a fraction of the cost of the total calculation.

Sensitivity analysis enables the investigator to help predict the variation of a parameter vector α on the dependent variables in the problem. The vector α may contain rate constants, mass flux fractions, etc.. In particular, the quantities of interest are the first order sensitivity coefficients $\partial z/\partial \alpha_k$, $k = 1, 2, \ldots, N$. The appropriate equations for these quantities can be derived by differentiating (3.2) with respect to α_k. We have

$$\frac{d}{d\alpha_k}\left(F(z;\alpha)\right) = \frac{\partial F}{\partial z}\frac{\partial z}{\partial \alpha_k} + \frac{\partial F}{\partial \alpha_k} = 0, \quad k = 1, 2, \ldots, N. \tag{3.8}$$

Upon recalling that the Jacobian matrix $J = \partial F/\partial z$, we can re-arrange (3.8) in the form

$$J\frac{\partial z}{\partial \alpha_k} = -\frac{\partial F}{\partial \alpha_k}, \quad k = 1, 2, \ldots, N. \tag{3.9}$$

We remark that although (3.9) can be solved at any step of the Newton iteration and at any level of grid refinement, we ordinarily solve the sensitivity equations on the finest grid with the last Jacobian formed. In practice, it is only at this stage of the calculation that the numerical approximation to J can represent adequately the analytic Jacobian. In addition, since the Jacobian matrix on the final grid is already formed and factored, each sensitivty coefficient can be calculated by performing a back substitution at the cost of an additional $5K^2M$ additions and multiplications (the cost of factoring the entire Jacobian is $7K^3M/3$ additions and multiplications).

4. Numerical Results

We have applied the algorithm discussed in the previous section to a variety of flames (see, e. g., [10-12,14-15,17]). In this section we consider application of the method to an atmospheric pressure ozone-oxygen premixed (adiabatic) flame and an atmospheric pressure methane-air counterflow diffusion flame. In both calculations we illustrate temperature and species profiles as a function of the independent coordinates and, in the ozone flame, we also calculate the adiabatic flame speed as a function of the percent of ozone in the incoming fuel-oxidizer mixture.

Premixed Ozone-Oxygen Flame One of the simplest premixed laminar flames is the ozone-oxygen flame. The reaction mechanism for this flame is sufficiently well understood that we can eliminate discussion of the effect of variations in the reaction mechanism on our results. We list the reactions and the appropriate rate constants in Table 4.1. In this flame we calculate temperature and species profiles and the mass flow rate (the adiabatic flame speed). We point out that the calculation of flame speeds as a function of the fuel-oxidizer ratio is particularly well suited to our boundary value method. We implement a continuation procedure in which the solution of the governing equations for one value of the fuel-oxidizer ratio provides an excellent guess for another problem with a different ratio. In addition, the adaptive gridding procedure automatically adjusts the number of grid points needed to maintain a specified degree of accuracy. If the location of the flame zone shifts as the fuel-oxidizer ratio is adjusted, the mesh adjusts itself accordingly.

TABLE 4.1

Reaction Mechanism Rate Coefficients In The Form $k_f = AT^{\beta}\exp(-E/RT)$. Units are moles, cubic centimeters, seconds, Kelvins and calories/mole.

	REACTION	A	β	E
1.	$O_3 + M \rightleftharpoons O_2 + O + M^a$	4.31E+14	0.000	22200.0
2.	$O + O_3 \rightleftharpoons O_2 + O_2$	1.14E+13	0.000	4600.0
3.	$O + O + M \rightleftharpoons O_2 + M^b$	1.38E+18	-1.000	340.0

[a] Third body efficiencies: $k_1(O_2) = .44k_1(O_3)$, $k_1(O) = .44k_1(O_3)$.

[b] Third body efficiencies: $k_3(O) = 3.6k_3(O_2)$, $k_3(O_3) = k_3(O_2)$.

The initial calculation was performed on an atmospheric pressure, 20% ozone flame with 10 nonuniform mesh intervals and a computational domain of 5.0 centimeters. With initial estimates for the species profiles given as cubic polynomial functions for the reactants and products, and Gaussian functions for the intermediates, 20 time iterations were performed on the initial grid with a time step of 1.0×10^{-5} seconds. The final solution for the 20% ozone flame was obtained after five grid refinements on a highly nonuniform grid with 52 points. Flame speeds for other fuel-oxidizer ratios were generated by the continuation method discussed above. Starting with the solution to the 20% ozone flame, the ozone concentration was increased in 5% increments until we reached 100%. The final solution used 101 grid points. A comparison of our computed flame speeds with the data compilation of Streng and Grosse [21] is shown in Figure 4.1.

In Figure 4.2 we illustrate calculated temperature and species profiles for a 60% ozone flame. As the figure indicates, we obtain "zero slope" temperature and species profiles at $x = 0$ and the adaptive gridding procedure is able to resolve effectively the regions of high spatial activity in both the temperature and the species profiles. In particular, for the 60%

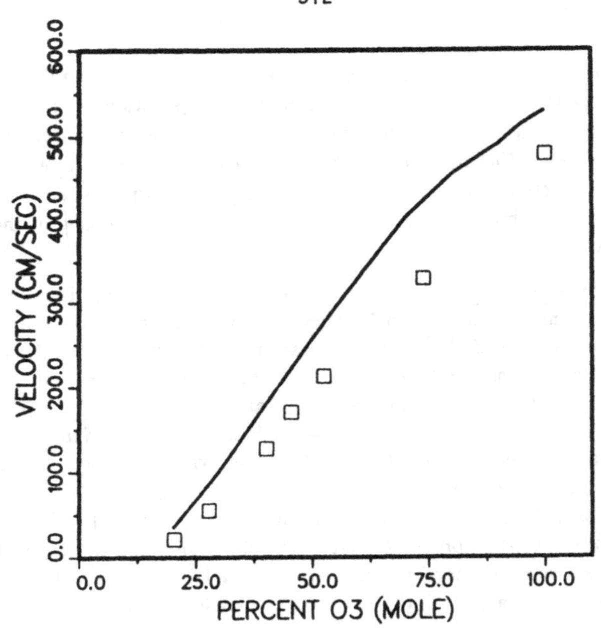

Figure 4.1 Comparison of computed atmospheric ozone-oxygen flame speeds (straight line) with the experimental data of Streng and Grosse [21].

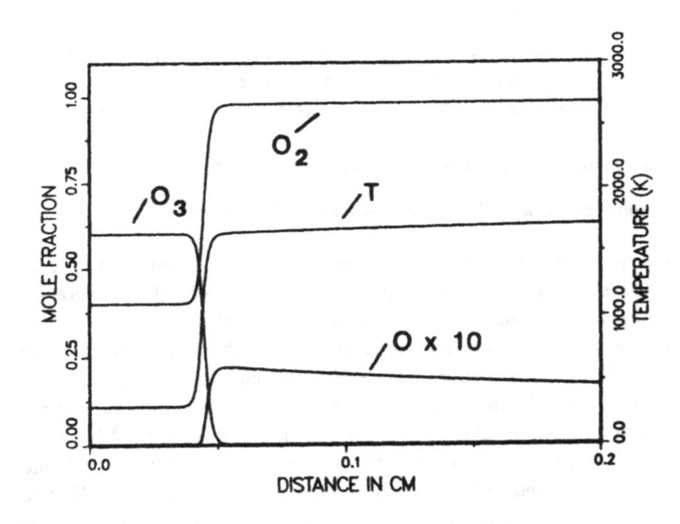

Figure 4.2 Computed temperature and species profiles for a 60% ozone-oxygen flame. In this calculation $x_f = 0.04$ cm. and $T_f = 350$K. The domain L was 5.0 cm., but the plot only shows the first 0.2 cm..

ozone calculation, 68 of the 101 grid points used were placed in the "flame zone" between $x = 0.0409$ and $x = 0.0644$ cm..

Methane-Air Diffusion Flame The flame we consider has been studied experimentally by Tsuji and Yamaoka [24] and detailed calculations for this flame have been carried out by a number of independent research groups (see Dixon-Lewis et al. [2]). Additional results, as well as the reaction mechanism used in the calculations in this paper, are reported in Miller et al. [12]. Counterflow diffusion flames are important in theories of turbulent non-premixed combustion. In particular, if the velocity gradient is varied, the scalar dissipation rate (a measure of strain in the flame) is changed and the effects of strain on the extinction properties of the flame can be investigated (see Peters [13]). Our purpose, however, is to illustrate the results of applying the time-dependent starting approach together with Newton's method to calculate temperature and species profiles for a methane-air counterflow diffusion flame with $a = 100$ sec^{-1}.

The calculation was performed on a computational domain with $0 \leq \eta \leq 10$. The initial grid contained 21 points and after several grid refinements a total of 66 grid points were used. The boundary data at $\eta = 0$ is given by

$$f'(0) = 0, \tag{4.1}$$

$$V(0) = 1.5, \tag{4.2}$$

$$\epsilon_{CH_4} = 1, \quad \epsilon_{k \neq CH_4} = 0, \tag{4.3}$$

$$T = 332.16\text{K}, \tag{4.4}$$

and at $\eta = 10$ by

$$f' = 1, \tag{4.5}$$

$$X_{O_2} = 0.21, \quad X_{N_2} = 0.79, \quad X_{k \neq O_2, N_2} = 0, \tag{4.6}$$

$$T = 283.16\text{K}, \tag{4.7}$$

where X_k denotes the mole fraction of the kth specie.

In Figure 4.3 we illustrate a comparison between the experimental temperature data of Tsuji and Yamaoka [24] and our calculated values. Our predicted temperature peak is more than 200K higher than their data. The discrepancy between the two is most likely due to the lack of a radiation correction in their thermocouple measurements. We also illustrate a constant-enthalpy, constant-pressure equilibrium temperature profile. In Figures 4.4 and 4.5 we illustrate several calculated species profiles together with the experimental data reported in [24]. In most cases the agreement is quite good. Finally, in Figure 4.6 we plot the peak flame temperature as a function of $1/a$. As the value of a is increased, the fluid dynamic residence time decreases and the maximum temperature drops until we are no longer able to compute a steady-state solution. This is the extinction limit for the flame. Experimentally the extinction limit for this flame occurs at approximately $a = 330$ sec^{-1} [25]. We compute a stable flame at $a = 330$ sec $^{-1}$ but the flame extinguishes at $a > 350$ sec^{-1}.

This work was supported by the DOE Office of Basic Energy Sciences.
[1]Department of Mechanical Engineering, Yale University, New Haven, CT.
[2]Combustion Chemistry Division, Sandia National Laboratories, Livermore, CA.
[3]Applied Mathematics Division, Sandia National Laboratories, Livermore, CA.

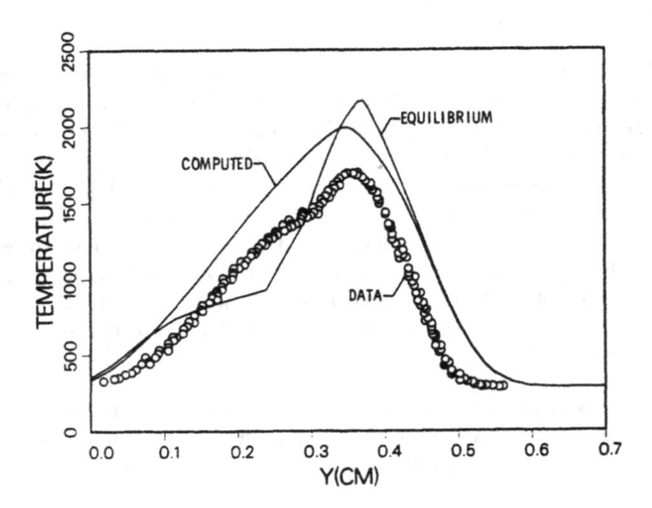

Figure 4.3 Comparison of calculated and experimental temperature profiles for the methane-air counterflow diffusion flame.

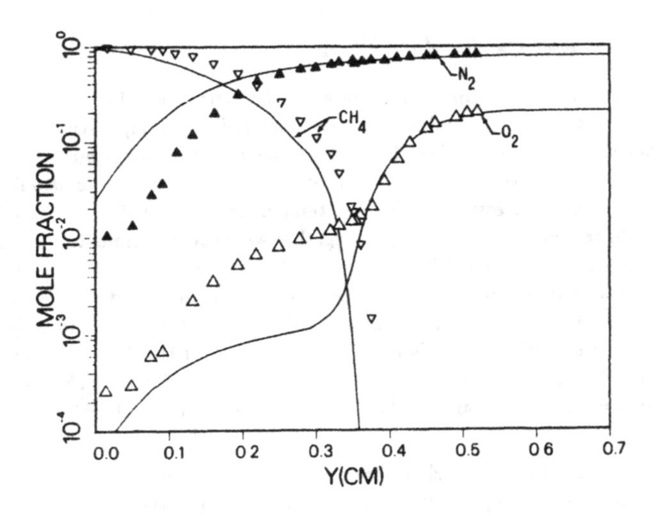

Figure 4.4 Illustration of calculated and experimental species (reactant) profiles for the methane-air counterflow diffusion flame.

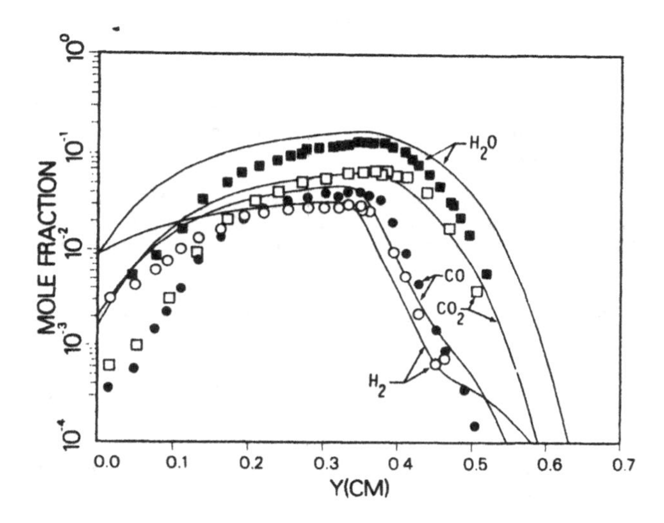

Figure 4.5 Illustration of calculated and experimental species profiles (products) for the methane-air counterflow diffusion flame.

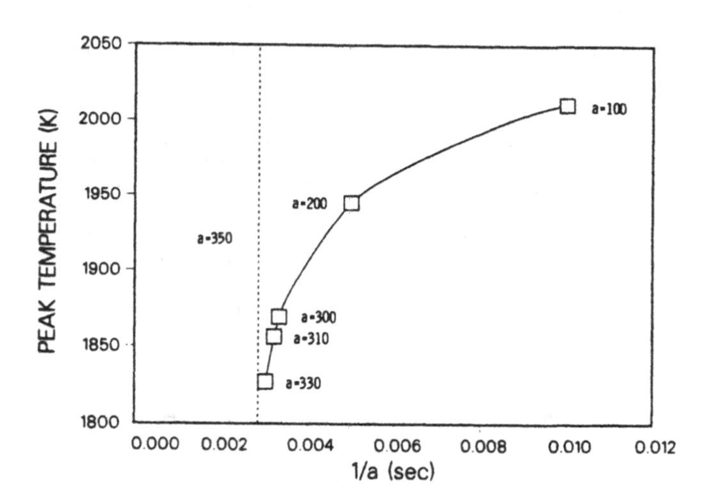

Figure 4.6 Illustration of the peak temperature versus $1/a$. As $1/a$ decreases, the residence time decreases and the maximum temperature drops until a steady-state solution no longer exists.

REFERENCES

[1]. Dixon-Lewis, G., "Flame Structure and Flame Reaction Kinetics I. Solution of Conservation Equations and Application to Rich Hydrogen-Oxygen Flames," *Proc. Roy. Soc. London*, **298A** (1967), p. 495.

[2]. Dixon-Lewis, G., Fukutani, S., Miller, J. A., Peters, N. and Warnatz, J., "Calculation of the Structure and Extinction Limit of a Methane-Air Counterflow Diffusion Flame in the Forward Stagnation Region of a Porous Cylinder," to be published in the Twentieth Symposium (International) on Combustion.

[3]. Frank, P. M., Introduction to System Sensitivity Theory, Academic Press, New York, (1978).

[4]. Hindmarsh, A. C., "Solution of Block-Tridiagonal Systems of Linear Equations," Lawrence Livermore Laboratory Report UCID-30150, (1977).

[5]. Hirschfelder, J. O. and Curtiss, C. F., "The Theory of Flame Propagation," *J. Chem. Phys.*, **17** (1949), p. 1076.

[6]. Isaacson, E. and Keller, H. B., Analysis of Numerical Methods, John Wiley and Sons, Inc., New York, (1972).

[7]. Kee, R. J., Miller, J. A. and Jefferson, T. H., "CHEMKIN: A General-Purpose, Transportable, Fortran Chemical Kinetics Code Package," Sandia National Laboratories Report, SAND80-8003, (1980).

[8]. Kee, R. J., Warnatz, J. and Miller, J. A., "A Fortran Computer Code Package for the Evaluation of Gas-Phase Viscosities, Conductivities, and Diffusion Coefficients," Sandia National Laboratories Report, SAND83-8209, (1983).

[9]. Margolis, S. B., "Time-Dependent Solution of a Premixed Laminar Flame," *J. Comp. Phys.*, **27** (1978), p. 410.

[10]. Miller, J. A., Mitchell, R. E., Smooke, M. D. and Kee, R. J., "Toward a Comprehensive Chemical Kinetic Mechanism for the Oxidation of Acetylene: Comparison of Model Predictions with Results from Flame and Shock Tube Experiments," Nineteenth Symposium (International) on Combustion, Reinhold, New York, (1982) p. 181.

[11]. Miller, J. A., Smooke, M. D., Green, R. M. and Kee, R. J., "Kinetic Modeling of the Oxidation of Ammonia in Flames," *Comb. Sci. and Tech.*, **34** (1983), p. 149.

[12]. Miller, J. A., Kee, R. J., Smooke, M. D. and Grcar, J. F., "The Computation of the Structure and Extinction Limit of a Methane-Air Stagnation Point Diffusion Flame," paper presented at the 1984 Spring Meeting of the Western States Section of the Combustion Institute, Boulder, CO..

[13]. Peters, N., "Laminar Diffusion Flamelet Models in Non-Premixed Turbulent Combustion," to be published in *Progress in Energy and Combustion Science*.

[14]. Smooke, M. D., "Numerical Solution of Burner Stabilized Premixed Laminar Flames by Boundary Value Methods," *J. Comp. Phys.*, **48** (1982), p. 72.

[15]. Smooke, M. D., Miller, J. A. and Kee, R. J., "Numerical Solution of Burner Stabilized Premixed Laminar Flames by an Efficient Boundary Value Method," Numerical Methods in Laminar Flame Propagation, N. Peters and J. Warnatz (Eds.), Friedr. Vieweg and Sohn, Wiesbaden, (1982).

[16]. Smooke, M. D., Miller, J. A. and Kee, R. J., "On the Use of Adaptive Grids in Numerically Calculating Adiabatic Flame Speeds," Numerical Methods in Laminar Flame Propagation, N. Peters and J. Warnatz (Eds.), Friedr. Vieweg and Sohn, Wiesbaden, (1982).

[17]. Smooke, M. D., Miller, J. A. and Kee, R. J., "Determination of Adiabatic Flame Speeds by Boundary Value Methods," *Comb. Sci. and Tech.*, **34** (1983), p. 79.

[18]. Smooke, M. D., "An Error Estimate for the Modified Newton Method with Applications to the Solution of Nonlinear Two-Point Boundary Value Problems," *J. Opt. Theory and Appl.*, **39** (1983), p. 489.

[19]. Smooke, M. D. and Mattheij, R. M. M., "On the Solution of Nonlinear Two-Point Boundary Value Problems on Successively Refined Grids," to appear.

[20]. Spalding, D. B., "The Theory of Flame Phenomena with a Chain Reaction," *Phil. Trans. Roy. Soc. London*, **249A** (1956), p. 1.

[21]. Streng, A. G. and Grosse, A. V., "The Oxygen to Ozone Flame," Sixth Symposium (International) on Combustion, Reinhold, New York, (1957), p. 264.

[22]. Tsuji, H. and Yamaoka, I., "The Counterflow Diffusion Flame in the Forward Stagnation Region of a Porous Cylinder," Eleventh Symposium (International) on Combustion, Reinhold, New York, (1967), p. 979.

[23]. Tsuji, H. and Yamaoka, I., "The Structure of Counterflow Diffusion Flames in the Forward Stagnation Region of a Porous Cylinder," Twelfth Symposium (International) on Combustion, Reinhold, New York, (1969), p. 997.

[24]. Tsuji, H. and Yamaoka, I., "Structure Analysis of Counterflow Diffusion Flames in the Forward Stagnation Region of a Porous Cylinder," Thirteenth Symposium (International) on Combustion, Reinhold, New York, (1971), p. 723.

[25]. Tsuji, H., "Counterflow Diffusion Flames," *Progress in Energy and Comb.*, **8**, (1982), p. 93.

[26]. Warnatz, J., "Calculation of the Structure of Laminar Flat Flames III; Structure of Burner-Stabilized Hydrogen-Oxygen and Hydrogen-Fluorine Flames," *Ber. Bunsenges. Phys. Chem.*, **82**, (1978), p. 834.

PROGRESS IN SCIENTIFIC COMPUTING
ALREADY PUBLISHED

PSC 1 The Finite Element Method in Thin Shell Theory: Application to Arch Dam Simulations
M. Bernadou, M.M. Boisserie
ISBN 3-7643-3070-8

PSC 2 Numerical Treatment of Inverse Problems in Differential and Integral Equations
P. Deuflhard, E. Hairer, editors
ISBN 3-7643-3125-9

PSC 3 Lanczos Algorithms for Large Symmetric Eigenvalue Computations Vol. 1, Theory
Jane K. Cullum, Ralph A. Willoughby
ISBN 0-8176-3058-9
ISBN 3-7643-3058-9

PSC 4 Lanczos Algorithms for Large Symmetric Eigenvalue Computations Vol. 2, Programs
Jane K. Cullum, Ralph A. Willoughby
ISBN 0-8176-3294-8
ISBN 3-7643-3294-8
ISBN for two-volume set 0-8176-3295-6, 3-7643-3295-6